Because of You,
I Lost Everything!

*True Stories Told from the Eyes
of a Missionary Kid*

Rev. Jim H. Darnell, Jr.

Cover by: Dawn Darnell

Cataloging in Publication Data

Because of You, I Lost Everything!

Autobiography

ISBN: 979-8839663510

Independently published

July 2022

Cover by: Dawn Darnell

Other books by Jim H. Darnell, Jr

Lightning Butterflies

Table of Contents

3 For I say through the grace that was given me, to every man who is among you, not to think of himself more highly than he ought to think; but to think reasonably, as God has apportioned to each person a measure of faith.

-Romans 12:3   (WEB)

Chapter 1

The Vision Catching Fire

Dad Is Gone

Dad died four years ago. It is strange how it works. When he died, he took a big piece of me with him, and he left a huge piece of himself in me. I feel so empty and so full at the same time.

God had spoken so many times, "This is the man that I honor – just watch and see." I watched and I saw. I saw so many wonders that I decided that nothing would be better than to be like him.

James and Jerlene Darnell (dad and mom) were international missionaries in North America, Africa, and South America whose passion was to start churches. As a little child, I was in awe when watching the Spirit of God grip with such fervor those who had just heard of Jesus. I did not know it wasn't commonplace for everyone to see such drastic spiritual transformations as a regular occurrence. I began noticing patterns of the ways God would bless and discovered that some things God loves to bless a lot more than others. Some paths are blazed with God's power and those who know how to find them are driven by the Spirit of God.

16 For the Lord himself will descend from heaven with a shout, with the voice of the archangel and with God's trumpet. The dead in Christ will rise first, 17 then we who are alive, who are left, will be caught up together with them in the clouds, to meet the Lord in the air. So, we will be with the Lord forever. 18 Therefore comfort one another with these words.

-1 Thessalonians 4:16-18 (WEB)

Nowhere Else to Go

Dad was preaching at one of the services in Ivory Coast, Africa. A man listened to the challenge – tell his family within 24 hours that he had chosen Jesus as his Savior. He knew that his family would reject him and cast him out. He had a choice between choosing His Maker or his family. Choosing God would have meant losing everything.

This man made his choice. He surrendered his life to God and took the challenge. He went home and told his family about his choice to serve the Living One. His family predictably spewed him out. This man was left with no way to care for his needs. He had no place to live. He had no way to supply the basics like food. Without mercy, he was destitute.

What he did next frightened me. He hunted down the man whose challenge cost him everything. This rejected man went to find dad. He was a massive man who could have easily done great bodily harm to dad, and I was probably 10 or 11 at the time. There, in the front yard, the encounter happened. He lost everything because of what my dad told him. This is what this man said. Excuse me, this man was not talking, he was yelling,

"I told my family how much I love them and of my love for Jesus. They were so mad at me that they gave me a choice. I could either renounce my new faith or I would be cast out of the family. I kept reminding my family how much I love them, and that they too could have the same touch of God in their lives as well. They cast me out. Here I am. I can't go back."

At this point, I could not tell if the man was grateful or furious with dad. Did this man expect dad to support him? Out there on the lawn, we stood. I was twenty feet away, which was close enough to hear, but far enough to run. I knew that whatever the man said next was going to have a tremendous impact on my life. With a finger jabbing toward dad's face, he screamed, "Because of you, I lost everything!"

This man, who needed so much, did not ask, or want anything from dad. Before, he had a place to stay and food to eat. He had the security and love of family. He once had his meaningful place in life passed down through the generations. Now he lost all that, yet he said, "Things are better now than they were before." Things were so much better now that he could not go back.

This man had come to understand things for real. While preaching, Dad had said, "You and God are a majority!" This man found out in all the rejection that he and God were the majority. God was all he needed. God so loved this man that nothing else mattered. The sloshing of God's love within drove this man to never return to the unsatisfying faith he had known. How could this man go back? When life was full, he was empty; now that life is empty, he was much fuller than he could ever imagine. God's love made standing up for Him so worth it. There was nothing more to say, this man had chosen the most incredible journey.

Dad just stood there looking at him. After a long pause, he continued, "Taking the 24-hour challenge was the best decision of my life." With that, the man spun around and marched off. We never saw or heard from that stranger again.

God and that man made a majority no matter what else happened.

*39 He who seeks his life will lose it, and he
who loses his life for my sake will find it*

-Matthew 10:39   (WEB)

Old Man

The Africans loved to call dad, "Old man!" They used the term "old man" for someone who is incredibly wise. When the Africans called dad "Old man," they were giving him the biggest compliment they could.

Dad grew up in the United States where being old is often looked down upon. When dad was called an old man, it meant that he used to be something that age had taken away. Dad always heard the American insult when he heard the term. Still, dad always responded appropriately because he also heard the compliment. Sometimes, men older than dad called him "Vieux!"

*32 "You shall rise up before the gray head
and honor the face of the elderly; and you shall
fear your God. I am Yahweh.*

-Leviticus 19:32   (WEB)

Chapter 2

Old Man

Old Man Style

Old man style is quite different than being called an old man. When I was young, I would hear old men ramble a story instead of getting to a point. These old men would frustrate me to no end. My mind was running, "Come 'on now!" or "Hurry up!" Since they were my elders, I had to wait it out until they finally got around to saying what they wanted to say.

The old men would talk about this, then they would talk about that. They kept jumping from one thing to another. Their direction seemed pointless, they would offer no outline and their way seemed unorganized, far too wordy, and always off-topic. They kept bringing in more and more subjects. Their style seemed terrible.

Suddenly, they would say their last sentence and stop. The last sentence took all these random ideas and tied them together. In an amazing twist, everything fit perfectly. Before I realized what happened, they poked me in the forehead with their point. I did not even realize that they were wrapping things up and they were done. When these master storytellers finished, I realized that in all my annoyance, I was left spell-bound.

The beauty of the style is that it was in story format instead of point outline format. The point outline format says what is going to be said, says what needs to be said, illustrates what was said, and summarizes what was said. The traditional point outline explains things well but leaves out the wonder that only a good story can bring.

If I just seem to wander and I leave you wondering, "What is the point?" I might just be using the old man style. I cannot promise that I will leave you amazed at the end, but some truths are better discovered than told. The outline of the book is by subjects, and the subject points build on themselves. Some of the subjects are more foundational and may not seem relevant at the time, but they are. The book is littered with stories of God touching lives. Also included are the reasons why mom and dad acted as they did, for God blessed their unconventional decisions with great power.

The Great Awakening

The stories are ever building on one another. Before Dad died, the Gospel went from unknown to flooding the country. These are the stories of the earliest of days of the stirring of the Spirit of God wherever my parents traveled.

The United States has not seen a Great Awakening since 1750. Many books have been written on the subject describing from historical records what makes such a sweeping movement of the Holy Spirit. Many have sought to be experts on the Spirit's most powerful moving within the United States and Europe.

The early days are gone, but God is still moving in Africa with great power. Africa's Great Awakening can still be seen. Even in Africa, the kind of blessing God poured upon mom and dad is unheard of. I have no desire to be a Great Awakening expert. Most of the stories at the beginning are stage settings for talking about the stories of the Spirit's wind.

Dad Was Mind-Boggling

I was twenty-five years old the last time I wandered on the mission field with mom and dad. Dad was telling me about his latest set of plans. I was blown away by the sheer magnitude of his planning. I have spent my life trying to become like him. After dedicating myself to trying to think as he did, dad's plans are still mind-boggling.

Dad thought very differently than everyone else I had met. Dad's approach to solving problems to this day makes him one of a kind. The uncommon person finds uncommon answers to common problems. If dad thought like the other experts, then dad would have been common. Dad did not think as they did, so his answers were always different.

Guaranteed Blessings

The overall focus of the book from mom's perspective is that God chose two very common people for God's work. From my perspective, I will try to express the kind of person God likes to bless. I will write about what kind of people mom and dad were. There are some things that God honors more than others. Some blessings are always guaranteed for certain steps of faith, the only thing that changes are the degrees to which the blessings can be seen openly. The blessings are always there for those who know how to find them. My parents found a way to stay amid God's abundance. Certain steps of faith always bring blessings from God. Enjoy the ride with me, as I invite you to step in the blessings with us.

6 Without faith it is impossible to be well pleasing to him, for he who comes to God must believe that he exists and that he is a rewarder of those who seek him.

~Hebrews 11:6 (WEB)

JIM H DARNELL JR

Chapter 3

Business Meetings

Church Business Meetings

Dad once told me that getting anything through a church business meeting was slow. Dad would talk to the leaders within the church. Dad's plans would overwhelm them. They would respond, "That would never work here." Before they ever tried it, they had already decided that dad's plan was no good. Keep in mind that dad had led most of these leaders to Christ personally. They patterned their faith after dad. They had chosen to give up everything they knew to be like dad. When dad would suggest something very strange to them, he was told that it would not work because he was an American and did not understand how things worked in Africa.

Dad had a definite style at these business meetings. Dad would not say much because the Africans needed to discover the truth for themselves. Someone at the meeting would get a spark of genius and say something. The crowd would like it. They would take turns each expressing how they thought the idea was wonderful according to the pecking order.

In the end, they wanted dad's approval. After everyone had spoken, it was dad's turn to speak because dad did not say anything. They all would turn to dad and wait for a response. Dad would have his serious look on. Dad knew that he could not tell them that their plan was not workable. In a typical dad style, he might give a very convincing nod. Dad had an incredible way of making people feel so incredibly special for a good decision. The good part of the plan dad would acknowledge, then dad would open their eyes. The Africans thought they had it all figured out when dad would bring up something else that they had never thought about. "Aaahh!" Their minds would go spinning again. The process would start all over again. Someone would say something good, and others would reinforce the idea according to the pecking order. Every man had to jump in and give his spill.

This process could take many hours like four, five, or even six hours. In the end, the Africans had come up with an outstanding plan that would meet their needs and avoid many problems. Their solution was just like what dad had suggested in the first place which they thought would not work.

Incredible People

Those who were drawn to Christianity were by far young men seeking their fortune. Many were unsettled young men who left the village seeking a better life. They had no skills for a trade, but they were eager for a chance. Many of these young people were at the bottom socially. They did not have their place in society, and they were searching.

If they had authority, then they would have had the experience to draw from. They had little to draw from to become spiritual giants nearly overnight. They became spiritual giants through being introduced to Jesus and developing plans to carry out their faith during the business meetings. These young people who lacked much education, trade skills, and experience as leaders were the one's mom and dad had invested their lives in.

These were the people that God gave to mom and dad so that they might reach the country for Christ. For dad's plan to reach the country with the Gospel, these people had to step up to become spiritual and social giants suddenly. There was no one else to draw from but these new very green converts without any experience.

The business meetings that inched along ever so slowly for dad moved in a whirlwind for the Africans. The Africans would go from thinking about making a meager existence to grabbing hold of a plan on how to reach their country for Christ. God put amazing people in my parents' path – people who decided to give their all to Christ.

The Vision Catching Hold

Mom and dad never sought to be heroes. Dad's plan was for the whole country of the Ivory Coast to be won for Christ. They were just so overwhelmed by the hopelessness that they had to have a plan that included everyone. Many could not or would not think like dad. Many times, mom and dad felt so lonely, because others could not grasp their way of thinking. How could a plan be adopted that considered some souls not worthy of salvation? The plan included all for the burden was felt for all. Dad spent countless hours nurturing missionaries and locals into a vision to reach all of the country for Christ. With the vision, dad passed on a plan. Despite much opposition, countless dedicated their lives to the vision and plan set before them. Little did they know at the time, but the effects were to dramatically touch far more than just Ivory Coast.

5 Listen, my beloved brothers. Didn't God choose those who are poor in this world to be rich in faith, and heirs of the Kingdom which he promised to those who love him?

- James 2:5 (WEB)

JIM H DARNELL JR

Chapter 4

The Younger Years

God's Touch on mom and dad

Dad grew up tough. During his era, there was no greater shame for a woman than to be divorced. Dad grew up under that shame. As dad once said, "I grew up on the wrong side of the tracks, and we were the poorest of the poor." There was a time when the railroad tracks separated the rich, educated, and influential from the poor, uneducated, and bossed around. His mom, Mandy Jane, was among the least significant women in the community.

Things were worse than insignificant. The most helpless and vulnerable people were bullied. Dad was bullied a lot. His brothers were gone, and his mom was busy working. Dad was at the mercy of the local punks. Dad lived life in the minus.

Along the way, dad made peace with God. Two things happened: on the outside, dad was harassed a lot, but on the inside, the fanfare of heaven resounded. Somehow, it was like dad was God's hero. God thrilled dad beyond belief. The Bible calls it love, joy, and peace. The creator of the universe was jumping up and down inside of dad with overwhelming approval. Do not ask me how or why the greatest there ever was adores the least, but God adores dad. Dad appeared like a loser on the outside and a winner on the inside.

11 Most certainly I tell you, among those who are born of women there has not arisen anyone greater than John the Baptizer; yet he who is least in the Kingdom of Heaven is greater than he.

-Matthew 11:11 (WEB)

God's Touch on Mom

Grandma, my mother's mom, was a harsh woman. I remember when I was five, she growled at me something fierce. I was in the wrong, but I did not need such a harsh reprimand. Our mom grew up under extreme criticism for not living up to expectations.

They were poor people who had little. Mom was not always around a church or anyone who would offer her nurture. It was common for people who lived in the neighborhood to grow up with a harsh tongue and full of bitterness. It was the way of life for those who were struggling just to get by.

Mom did not rebel, she just leaned on God's gentleness. Mom, just like dad, found that the wonders of heaven freely poured into the helpless who opened up to God. In the middle of so much harshness, Mom grew up gentle.

Part of the wonders of the story is that grandma blossomed into a gentlewoman. The rough edges disappeared, and she became someone I admired greatly before her death. Mom's gentleness rubbed off on grandma. Grandma saw what mom had and wanted it for herself. God is so grand that He splashed all over mom with love, joy, and peace, only to bless grandma the same. God is amazing. Mom, dad, and grandma learned to stay in God's splash zone.

20 Behold, I stand at the door and knock. If anyone hears my voice and opens the door, then I will come into him, and will dine with him, and he with me.

-Revelation 3:20 (WEB)

Mom and Dad's Passion

Dad was never ensnared by dreams of grandeur. Mom and dad never wanted to say, "Look what I did!" As missionaries, they never had anything to prove. They did not need to show anyone that God moved powerfully within. The ministry was never about them.

Pride drives away the splash zones. Seeking to prove something or seeking grandeur dries up the thrill. For God's wonderful splashes to increase, one's self must become less so that God would become more.

Mom and Dad had suffered so much in their childhood and never wanted to return to the feelings of being insignificant. Being insignificant on the outside did not matter if they felt significant to God. Mom and dad chose to spend the rest of their lives in the splash zone.

Dad was a tough man. It did not matter how many would oppose him, he would not budge an inch in terror. Many times, dad stood alone against large numbers who aggressively opposed dad but did not shake him. Dad was terrified of one thing – dad was terrified of losing God's splash.

Mom and dad had made a choice – they were going to live in the splash zone. They chose to spend their days amazed by the incredible God. As long as mom and dad trembled at being something, their insides trembled in wonder.

8 To me, the very least of all saints, was this grace given, to preach to the Gentiles the unsearchable riches of Christ

-Ephesians 3:8 (WEB)

Mom and Dad's Calling

Dad knew from the time that he stepped foot in Africa that his calling was to win Ivory Coast for Christ. The calling was not a win-lose proposition. It did not matter if they succeeded in winning over the country. Success in numbers was not the point.

The focus was on the splashing of God's wonders. Mom and dad wanted Ivory Coast to be soaked in the wonders of Christ. The focus was on others experiencing God just like mom and dad did. Mom trusted God for a husband that would enable others to experience God the way she did. Mom chose to let dad lead. She believed in her husband and trusted that God would lead dad.

7 But Yahweh said to me, "Don't say, 'I am a child;' for you must go to whomever I send you, and you must say whatever I command you.

- Jeremiah 1:7 (WEB)

Dad loved Sports

Dad played a lot of sports growing up. Dad rarely talked about being afraid. During his senior year in high school, he became afraid. He was his high school's running back. That year, WWII ended, and many soldiers returned from war. Many went off to war without finishing high school. They left as boys and returned as men. Many had developed physically during the war. It was not just his team that had a lot of returnees. Dad practiced against and played games against soldiers.

Dad also loved baseball. In high school, dad played shortstop. Dad batted over .400. After high school, dad continued to practice and play baseball. Dad played on several club teams. Eventually, dad moved to catcher. Dad progressed toward better teams. Dad always batted over .400. There was this semi-pro team that dad wished to join. The coach kept stalling about giving dad a chance. One day, they were short a player. Dad was called to fill in for just one game. Dad had three hits in that game. Dad outplayed the team. The coach immediately invited dad to join the team. About that time, a pro baseball team invited dad to try out. Sports were finally opening up for dad. For the longest time, dad wanted to be a ballplayer.

Mandy Jane Darnell

Mandy Jane knew her destiny. Her life's passion was to raise a pastor. God had engraved upon her heart that the greatest thing she would do in life was to raise a minister. She gave birth to a son. Paul had the perfect personality to be a pastor. Mandy Jane was convinced that Paul was the appointed one to become the pastor. Her son left home at about the age of 17. He left to pursue a successful business career. He never became a pastor.

When Paul was about six years old, Mandy Jane gave birth to a second son. Clarence also showed much promise. Mandy Jane was just certain that Clarence would enter the ministry. He left home at about the age of 17 as well. He was eager to make his way and had no desire for the ministry.

Mandy Jane's life was about to fall apart, but there was still time for destiny. When Paul was about 12 and Clarence about six, she gave birth to one more child. This child they named James. Mandy Jane was a stay-at-home mom for the two oldest children, but when James was about five, her world crashed.

Mandy Jane and Clarence Sr. divorced. He left, and she never remarried. Dad no longer had a stay-at-home mom. Mandy Jane worked non-stop for she was determined to put food on the table and provide a place to live. She no longer had the opportunity to raise a pastor. She was too busy, and James was left by himself. Dad remembered her working so hard. He remembers her rushing so much that she would run rather than walk as she did the laundry of richer folks. She just raced between tasks. Because she did not have time for James, she became more certain that Clarence was the one.

James was simply different. She just could not see him as the one who would complete destiny. I can imagine her heartbreak as she realized that Paul, her oldest, wanted nothing to do with the ministry. She had to wait another six years for Clarence to become old enough to decide what he desired to do. When both rejected the ministry, all she had left was James. Her life was rough. She bore the disgrace of divorce. She had to work all the time and had no time for herself or to raise her destiny. Still, her life had a keen sense of destiny because she felt ordained by God to raise a pastor.

If someone was going to complete her life's destiny; it had to be James, but James was not seeking the ministry either. She had high hopes that were ignored. James left the house after high school and enrolled at the University of Oklahoma. James did not take his studies seriously and flunked out after his first semester. Finally, James woke up. He knew that if he was going to be the one who completed Mama's destiny, he had to choose right then. James enrolled in Oklahoma Baptist University where he studied music and psychology. James had chosen the ministry.

Mandy Jane was never able to see all James was to become. She knew this; James was destiny. She could not prepare him by being the mother she had been for the other two. Everything that James became, passed through his mother. James chose the ministry because God had spoken destiny through his mother.

It has always saddened me that she did not see all that James became. To find comfort, I had to look beyond the obvious. Her life became a mess. She lived in dishonor without ever the honor of raising a minister that God would powerfully place his hand upon. She did not see much with her eyes. She did witness the early days of his ministry. At the same time, her faith eyes were wide open. Every mundane moment was accepted with great honor. She was to raise a pastor that God had appointed.

On the other side where earth sees no more, there is a grand lady who saw her son become more than just a minister. He became a living legend. About 4 years ago, she had the opportunity to welcome her son into eternity. There, God face to face could tell both of them, "Well done, you completed your destiny! Your faithfulness changed so many lives."

I was about 5 when Mandy Jane died. I don't remember her. I have no recollections. I have nothing on my own to say that she was a great lady. My parents have often asked me if I remember sitting on her lap shortly before she died of cancer. All I can say is "No!" It was not until later in life that I had an appreciation for what she did. She suffered so much for so little success. She lived in the shadows of failure. God only appointed one standard of success; she was going to raise a minister. There is a side of me that sees that she could have suffered less and had a better life, but eternity would not be the same. The greatness of her suffering is the measuring stick by which God will proclaim throughout eternity, "Look how much she loves me!"

No Pro Ball

Dad received the invitation to try out for a professional baseball team. When the time came for his dreams to come true, dad was no longer interested. The calling to become a pastor had become his own. He preferred to be a minister rather than seek fame and success.

JIM H DARNELL JR

Chapter 5

The Birth of a Church

Starting a New Work

In Ivory Coast, the fields were ripe everywhere. Studies could have been done to support this belief. Dad in his way studied the field. Dad knew of many wonderful places to start a church. Still, it is not studies that drove dad to start churches in new areas. Dad became burdened for communities and villages. Dad's heart would burn so mightily for these pockets of people that did not know the splashing of the wonders of God.

It would grieve dad that the fields were ripe and there was no one to harvest. Many would believe if only someone would go. Dad had a heart for others whom he saw were helplessly searching for God. It was the community with the heaviest burden upon dad's heart that drove him to start new work.

37 Then he said to his disciples, "The harvest indeed is plentiful, but the laborers are few. 38 Pray therefore that the Lord of the harvest will send out laborers into his harvest."

—Matthew 9:37, 38 (WEB)

Greeting Prospects

Before mom and dad ever had a service, they would always greet the people. Dad had a system by which he visited church prospects. He would always go find the leaders of a community first. He would tell them that we were planning to have a service here and that he wanted them to know first. They would always smile and were flattered by the courtesy.

Dad was always amazing to watch. Dad would always show such excitement for everyone he would meet. I can see him now holding his hand way off to the side and spinning his arm around to greet someone with great enthusiasm. Dad would smile or laugh and say it is so good to meet you. Dad would invite this person to go along with him. Dad would ask, "Do you have anyone you would like me to meet?" Dad would carry this ever-growing crowd around with him. Dad would not stop until he had met everyone in the village.

During his visit to start new churches, dad would greet first the leaders. Dad would not stop greeting until he had greeted everyone. Dad would make each one feel so special, and he would bring them into his entourage.

40 "The King will answer them, 'Most certainly
I tell you, because you did it to one of the least
of these my brothers, you did it to me.'

-Matthew 25:40 (WEB)

Showing a Movie

Once dad had the people gathered up, he would often show a movie. Many of the Africans had never seen a movie. It was a big treat for them. Dad had planned his visit so that by the time he finished greeting everyone, dark would have come.

Dad would remove a generator from the car. Dad never knew if he would have access to electricity. Besides, persecution often came. Dad would not want someone who had never heard the Gospel be persecuted for having shared his electricity.

Because of the roaring of the generator, dad would place it a long way away from the projection site. Dad would need several extension cords to cover the distance. He would hang up a string of hand-made lights. An extension cord was spliced and reattached ever so often to add in a light fixture. Even the sets of lights needed extra extension cords. These light cords were often hung from trees or on buildings. Splitters were needed so that many things could be plugged into one long extension cord. This jungle of electrical ropes was a wonder that sometimes took a sinister twist.

The movie projector was reel to reel. The movie tape was inserted into the track within the projector with a couple of feet of tape hanging out. The extra tape was wound on the empty reel. The tape was nothing more than a set of tiny pictures. On each side of the pictures was a set of holes so that the teeth in the projector could guide the picture. Somehow, the projector would stutter and pause the picture just briefly in front of the powerful lamp. As the movie flowed, the empty reel would gradually fill up and the full reel would gradually empty.

The projector worked great sometimes. If the movie strip came out of its tracks, then the lamp would only focus on one slide. If that happened, we would try to snap off the projector quickly. The heat from the lamp was immense. Sometimes, we would not be fast enough, and the lamp would melt the tape. We would then have to cut out the damaged pictures and splice the tape. The patch would usually hold up fine, but there would always be a little glitch on the screen.

Sometimes, dad faced persecution. That persecution would make me afraid. Dad's set-up had many vulnerable points. The generator was attached to many different extension cords. Someone would unplug one of the extension cords and everything would go dark, and the movie would stop. The only people that could be seen were white. Black skin in the dark is really hard to see. When the interruptions would not quit, what would they try next? Dad was not about to back down. He would find the source of sabotage and plug things back together. He would return to the projector and continue the movie. That was Saturday.

The Children's Choice

Ever since the beginning of the mission work, mom and dad gave my sisters and me a choice: Do we want to stay in the mission field? They made it clear that the mission work did not just belong to the adults, but to us children as well. They made it clear that we had a say so in the matter. I do not know what our say so meant, because Patty, Julia, and I always chose to support mom and dad. Being on the mission field was a choice we children chose to make.

Since being on the mission field was a choice, we believed that every day we were there, we were contributing to the ministry. I don't think mom and dad would have stayed on the field without the support of the children. We were not just tag-alongs; we were a vital part of the work. We children did our part in helping mom and dad be successful.

18 Yahweh drove out from before us all the peoples, even the Amorites who lived in the land. Therefore, we also will serve Yahweh; for he is our God."

- Joshua 24:18 (WEB)

Sunday before Church

The main ritual for the first service was always the same. Patty, Julia, and I would load our small station wagon down with lots of chairs. Mom and dad did not participate in this ritual because it was the children's job. Usually, forthe first service, we loaded 25-30 chairs.

On the way to the service, the chairs were quite annoying. We would fill the hatch up. The luggage rack above would be loaded down with chairs as well. The station wagon was three benches seating three on each bench for a total possible of nine. We would often seat three in the first seat. The back seat would fold down, and that area will be filled with chairs. Where the third person would sit in the middle row, we would stack chairs as well. The chairs were so wide that they would gobble up much of the space in the other two seats. All the sitting room that remained in the middle row was for two small children's frames. We felt pinched between the rolling stack of chairs and the door.

Being all cramped up for a long journey was not the big deal. Back in the day, cars did have air-conditioners, but they were almost useless. On a short trip across the city, the car would begin to cool off when we arrived. With five hot bodies in the car, the heat would stay unbearable for a long time. The bodies up front shielded the cool air from us in the back. The massive number of chairs killed any possible breeze. Remember we lived in the tropics. The humidity was always very high just like the temperature.

Arrival at the First Service

When we arrived, no one would be there. No one dared to get there before the missionaries. The Africans cared much more about event time than they did about a clock. When we showed up, they would leave their houses and come to us.

The first step would be to unload the car. The Africans very quickly unloaded the vehicle and set up the chairs. On each chair was placed a little hymnal. During the set-up time, people would continue to flow in. We would all start the tasks. The arrivals would see what we were doing and pitched in. The work was always completed very quickly.

2 Blessed be Yahweh's name, from this time forward and forever more.

-Psalms 113:2 (WEB)

Greeting the People

Mom and Dad would be busy greeting people. I would make sure to greet everyone as well. They all knew the proper European handshake. Very often the men would be incredible hulks. They did everything by hand, and so they grew massively strong. This was a big deal to a small boy with little hands. Not only did adults have way too big of hands, but some of the hands were super extra-wide due to their massive strength. The Africans believed in a firm handshake. The Africans were also very eager to see us. The combination of an adult hand, super-powerful, extra-wide, and a super firm grip due to sheer excitement was crushing to little hands. I had mastered the art of protecting my hand. I tried my hardest to reach my fingers to the other edge of the hand so that I could curl my hand within theirs. If I could keep mine curled, then the shake would not hurt. Sometimes, nothing would help, and my hand would be squeezed down to nothing. During the squeezing, my knuckles would be rolled together creating momentarily stunning pain. The only thing I could do is get ready to put my hand out again when we left. It was my calling – it was what I could do.

Patty's Ministry

Patty was a darling. Most had never seen whites before. Patty was the ultimate white – she was blond. Her blond hair drew the attraction of many young men. Patty drew in large crowds, and many were saved because of her. I do not know how Patty did it, but not only did she bring in the crowds, she entertained them as well. Patty was very observant of those who were the outcasts socially. She made sure that she included them.

15 Most certainly I tell you, whoever will not receive God's Kingdom like a little child, he will in no way enter into it."

-Mark 10:15 (WEB)

Marriage Proposals

Many of the young men thought that Patty was a great catch and were eager to marry her. Africans believed in being bold for being brave was honored. Many times, these young African men would propose only after having just met her.

Every time we went somewhere new, we would ask Patty, "How many guys proposed to you today?" On a really low day, Patty would say, "Only one!" Or "Only two!" If it were an average day, she would say, "Three" or "Four." Patty would get excited when she would have more marriage proposals. One day she glowed and said, "Seven!"

The Pinchings

One of the most unpleasant things was the pinchings. Africans had little understanding of whites. Their hair colors were different. I had brown hair with red tints. Julia had dark blond hair. Mom's hair was brown, and dad had black hair. Patty had the ultimate hair attraction; she was a light blonde. The Africans did not know what to expect. They would walk up to us and brush past us. In the passing, they would pinch us. The attempt was not to hurt but to see if white skin felt the same as black skin. After they would get their fill of pinching us, it would stop. There would often be so many pinchings at one time that it was impossible to tell who was pinching. I guess that they did not dare to pinch dad for he was an intimidating man.

Julia's Ministry

Julia is six years younger than I am. After Julia turned two, she was granted permission to walk around in the gathering of people trying to gain her attention. Mom and dad were brave to set her down and let her walk during such a chaotic event. Julia stayed in the midst of us, but she had her own space as well. Julia was really brave. She understood her part as well in caring about people. She went beyond the toddler fears of strangers and welcomed people. Julia would allow countless strangers to surround her. Seeing Julia amid the throngs of people who surrounded her often became difficult. She let people huddle around her and greet her. What a glorious impact she made on the kingdom of God.

14 But when Jesus saw it, he was moved with indignation, and said to them, "Allow the little children to come to me! Don't forbid them, for God's Kingdom belongs to such as these.

-Mark 10:14 (WEB)

The Song Service

I always felt awkward during the first hymn service. Most could not carry a tune. Even if they could carry a tune, no one knew any of the songs. Everyone held a hymnal proudly. Many could not even read, but that did not keep them from holding a hymnal up high. Some of those who could not read were obvious. They made quite a sight when they proudly sang from an *upside-down* hymnal. They did not even know what they were singing. They had no understanding of Christianity or what the words meant. Still, the Spirit of God was touching lives.

*92 It is a good thing to give thanks to Yahweh,
to sing praises to your name, Most High,*

-Psalms 92:1 (WEB)

God's Vision for Dad

Two things stirred Dad more than anything else. First was the incredible love God had showered on dad. God's love on the inside overcame the troubles on the outside. The second one has already been mentioned, but not yet explained. Dad had a vision for reaching Ivory Coast to Christ. God spoke to dad with such power. God spoke to dad and said, "I want to reach through you just as far as I can reach." This is such an incomprehensible prospect. How can the Creator and Owner of the universe hide Himself to reach through an insignificant person? Yet God's plan from the beginning was to become nothing to make others everything. God chose to make dad His everything.

God would stir my dad into a frenzied excitement. God was ecstatic about what He wanted to do through dad. Amazingly, anyone can live without catastrophic failure with the amount of excitement that God pumps into a soul. God placed within dad a calling that was bigger than life.

Love truly is supreme, not the calling. Those who chase the calling first lose their first love. Those who love the call see production as more important than people. It is terrible when those who hinder the calling are seen as obstacles to be conquered rather than people to be loved. When the calling is more important than loving people, bitterness, and all sorts of ugliness flare up within those who are consumed with the mission.

Mom and dad never lost sight of people. They were driven to the calling to love. The terrors of their childhood drove them to God's love. Mom and dad did not want to go back to the emptiness of terrors; rather, they saw the same helplessness in the people around them. God's calling upon my dad was very specific. "Let me reach through you as far as I can reach to win Ivory Coast." Dad was on a mission – God's mission.

The outcome of the calling is rarely clear. Eyes should be kept off the outcomes of life, but rather on Him. The calling makes the most ordinary moments extraordinary because God ordains every moment. God does not love mom and dad any more than He loves anyone else. The same rules apply to everyone. God desires to overwhelm everyone with His love and ordain every movement as on a mission for Him. The only way to understand the calling is to search out His love.

Love is supreme. God's love is the fuel for the calling. Dad believed it was his mission for Ivory Coast to come to Christ. Many churches were born, through the vision, the faithfulness of the children, and the eager reception by many Africans.

Chapter 6

The African Search for Meaning

The African Listeners

The Africans saw the world passing them by. Most Africans had very little and saw no opportunity to have a better life. Many African men saw village life as no way to live. They would rather be hungry with only the rags they wore and a mat to sleep on in a place that was not their own rather than live in the village. These were brave men willing to lose their security for a chance to have more.

Before we continue, here is a reminder that African village life did have a lot to offer. There were rules of respect that had to be followed. They were surrounded by family, and they had a way of life that sustained endless generations of Africans. With all the good, there was still enough missing to drive large numbers of men to the city.

Village religions were often terrifying. Africans understood that everything that happens in the natural world was because of the spirits. If something broke, then it was because the spirits wanted it to break. If someone fell and was hurt, it was because someone else has cursed them through the spirits. Every good and bad event was driven by the spirit world.

In the Western world, the focus is on natural causes. The world has been set in motion and all causes are strictly natural causes. If there is a serious car accident, then someone must be to blame. It is because of the drunk driver that the collision happened, and not because of the spirit world's involvement according to western culture.

Western culture is very quick to point out that spiritual involvement in everything is wrong. It is very correct that everything is simply cause and effect, or is it? Does God give and take away? Do demons seek to destroy? Western culture accurately sees everything as a cause and effect, but they are often blind to the spiritual causes and the certain effects that follow.

It would seem most logical that the most successful people on the face of the earth would say, "I did it!" and take for granted the God-given opportunities. Western culture has moved away from being deeply religious as prosperity has increased to say, "I don't need God, because I did it for myself!" Westerners will one day fully see the spiritual causes and effects.

African culture in Côte d'Ivoire does not have a way to reach God directly, they can only reach the spirits. Africans are all the time making gifts to the spirits to appease them. It was common to see a huge tin tub sitting beside the road full of the best fruits and vegetables all neatly wrapped in a cloth. Hungry Africans would pass by every day and would not touch it, because they did not want the wrath of the spirits. The Africans were all the time reaching out to the spirits to make life better.

Two common themes appeared as the Africans tried to appease the spirits. The more the Africans interacted with the spirits, the more frightened they became. Reaching out to spirits is very frightening. They found that contact with the spirits was not friendly. The first agenda of the spirits was to terrify. The peace and security that the Africans so desperately sought, instead brought greater insecurity and fear when they reached out to the spirits that spun evil.

The second theme is equally upsetting. The whole idea of honoring the spirits for some control in life is false. The more Africans tried to give to the spirits, the more the spirits took. Those who tried to be responsible and make peace with the spirits found themselves more controlled by the destructive spirits. Rather than controlling the spirits through gifts, the spirits forcefully controlled the giver. Africans have found out the spirits to whom they reached out were wickedly evil. Seeking peace and control by being nice to the spirits always brought fear, bondage, and ruthless control.

The Africans sought love and significance. They felt unloved as the opportunities of life rejected them. The Africans also longed for meaning in life. Sure, the Africans left the village to find their fortunes, but had they found the love and significance they were so longing for in the village, they would not have left to search for it. Instead of love and hope, village life brought spiritual fear and control without significance for a large number. Many Africans were ready for new life.

7 They shall no more sacrifice their sacrifices to the goat idols, after which they play the prostitute. This shall be a statute forever to them throughout their generations.'

-Leviticus 17:7 (WEB)

12 For our wrestling is not against flesh and blood, but against the principalities, against the powers, against the world's rulers of the darkness of this age, and against the spiritual forces of wickedness in the heavenly places.

-Ephesians 6:12 (WEB)

The Laborer's Curse

The Africans had repeatedly felt rejected. The rich and

powerful hired the African daily workers for next to nothing and pushed them to work harder and faster. I heard countless stories that when payday came around, the bosses would scream profanities at the workers. The supervisors yelled that they had never seen them before and were going to threaten the police on them. The workers spent a week paying their way to the job site. The extreme hard labor required much food. At the end of the week, the laborers had nothing but humiliation to show for their efforts and dreams.

4 Behold, the wages of the laborers who mowed your fields, which you have kept back by fraud, cry out, and the cries of those who reaped have entered into the ears of the Lord of Armies

- James 5:4 (WEB)

Chapter 7

Dad's Preaching Advice

Seminary Lectures

I remember eight years after leaving Africa when I went to seminary, and many times I heard the same thing. These professors were the most up-to-date on the most current trends in preaching. The professors drove it into us that the evangelist's job was to present the plan of salvation. The focus was to make the plan of salvation clear.

Very similarly, I heard the same thing expressed slightly differently. The most important thing that a minister can do is to lead the lost to Christ. They gave two reasons. The church needs new converts to survive. Salvation is the key to spending eternity in heaven rather than in hell. Seminary's focus was on salvation. While seminary professors and students had good points, dad had a unique way of thinking.

Dad's Preaching Passion

You would think that an evangelist like dad would center his preaching on salvation, but he did not. Dad focused on what he knew best. The sloshing of God's love within and the wonderful calling. Those are the two things that motivated dad.

The reason the people were so drawn to dad is that they were so convinced dad loved them. Before dad preached, he went around loving the people. The most important thing that could ever happen is to be loved. This American who chose to leave it all to live with those who had nothing earned much credibility. Because dad loved the Africans, they were ready to believe him.

Dad had a dream for the Africans. Dad dreamed of the Africans being overwhelmed by God's love. Dad wanted to pass along a vision through which the love could flow. Dad saw people that God highly loved, and ones He wished to make His hands and feet. Dad wished to give the Africans the same God that he had. Dad dreamed the Africans would be thrilled by God's love and be on a mission for the Creator.

Dad had a favorite phrase when he preached. Dad would declare with all his conviction, "God wants to reach through you as far as God can reach!" The most incredible thing happened, so many Africans realized that they had just found what they were looking for. The hole within the Africans could not be filled by ambition, reaching out to the spirits, or by anything else they tried. God told them, "I am eager to fill that hole as I reach through you and amaze you!"

Dad would pass along his dream while God was preaching in the hearts of the listeners with far greater power. God would preach themes to similar these. "I want to reach your family and your friends. All of Ivory Coast needs the hope and love I offer. I have called you to be the ones who tell them. If it is not you, then who? Unless you tell them, how will your family make peace with me? You are the ones whom I have chosen to tell them."

5 "Look among the nations, watch, and wonder marvelously; for I am working a work in your days, which you will not believe though it is told you.

—Habakkuk 1:5 (WEB)

40

Dad's Preaching Advice 1 of 3

Dad and I often played soccer with the Africans. I was a young boy who was excited about playing soccer. On one particular day, I had no grand aspirations, but dad did. I had just finished a very satisfying day playing with my friends and with dad. After we had finished playing, my friends left. The sun was about gone. This was the moment that dad wished to give me the secret to his unending passion. At the time, it was a pretty cool idea, but it was something that would grow deeper in meaning as the years flew by. Later, I looked back and realized that dad had waited for the right time to tell me his greatest secret to preaching. It was time to play the old man's game and change the subject – just kidding.

Dad knew that I wanted to preach, but I did not have the communication skills at the time to express myself very well. Thinking about preaching was pointless because a sermon had to be more than a sentence, and a sentence was all that I could say. On that day, dad's sermon was a sentence. A sentence that forever changed my life. I was thinking about soccer when dad said, "Preaching is giving the Spirit of God room to work!" That was all dad had to say, and from dad's perspective, that was the end.

I did continue the conversation. Dad often spoke powerfully for dad's words were very motivating. Dad spoke and people were ready to jump. How can a man say he does not do what he so obviously does? I asked dad, "So you don't try to convince the people of anything?" I was in shock. The greatest motivational speaker that I knew did not try to convince the people of anything. In preaching, dad chose to be the assistant. Dad's reply was, "In preaching, I must tell the truth, but the Spirit of God changes hearts."

When my time came as an adult to preach. Dad had lifted an incredible burden. If the people did not accept the words spoken, then God is to bear the burden of the people not responding. Preachers have never been responsible for the hearer's response.

A second impact was a total change in preaching focus. The sermon's focus is not on what humans need to do, but on what God seeks to do. Somehow, the most effective preachers do pass along the very picture of God's desired handiwork. It is an amazing thing when a preacher can so precisely describe what God is up to. Sermons need to have a heavenly perspective describing what God is longing to do at that particular time.

Dad was brave to reveal his secret of success. The best preaching is setting the stage for the Spirit to work miracles. Preachers are never responsible, nor do they get the credit for the outcome. Great preaching is presenting what the Spirit of the living God is up to. The Soul Shaker breaks the strongest souls, yet many of the broken still walk away with nothing but despair. The preacher shares in the joys and the sorrows of the sermon response, but the preacher never is responsible for the outcome. The preacher best sets the stage by being in tune with what the Spirit is ready to whisper. The effective preacher never claims success nor tries to figure out what the listeners are willing to hear. The genuine preacher passes God along.

12 For the word of God is living and active, and sharper than any two-edged sword, piercing even to the dividing of soul and spirit, of both joints and marrow, and is able to discern the thoughts and intentions of the heart. 13 There is no creature that is hidden from his sight, but all things are naked and laid open before the eyes of him to whom we must give an account.

-Hebrews 4:12, 13 (WEB)

Reaction to Salvation Messages

Often missionaries focused on the need for salvation. These were the messages that God gave many of the other missionaries. Messages are appointed for a specific time and event by the Almighty. Messages that have been God-honored throughout the ages. There is no better sermon than a call for Salvation if that is what God gives. Salvation sermons can be quite effective. Sometimes, a simple salvation message is exactly what someone needs. Some respond to simple salvation messages and become spiritual giants.

Purpose of Talking Salvation Strategies

The reality is that it seems that God blesses some people more than others. The goal here is to try to understand the mind of Christ so that we can enter into the greater blessing. Why would we want to be blessed less when it is perhaps possible to be blessed more? While it is important to know how to present Salvation, in a way we are not even talking about the presentation of the Gospel. The search here is – what can we do to unlock the Spirit coming with great power? If the Flow of Life is waiting on a cue, what is that cue?

Dad preached and people responded. Hearts were on fire and salvations happened. Lives were radically changed in a few moments. Dad's approach is different than any I have ever heard. When dad preached, God often fell upon them with incredible power. The majesty of the Almighty could not be seen, but this invisible power gripped the audience. Dad's style was very uncommon. I have wrestled countless times with the question, "Why did God so bless dad?" Many times, dad's presentation did not seem particularly good, yet the Spirit rushed in with great power. Was it because dad preached in a way that was more suitable to bless? Was it because God so loved dad? Was it just because? I don't want to answer that question but just wonder. It is probably some of all. However, there are some styles of preaching that more clearly reflect the moment that God has been waiting for.

The following section is comparing different ways of presenting salvation; yet in a way, we are not talking about salvation at all. As salvation is discussed, the goal is to find what God is passionate about. Just imagine that we can identify the themes that stir God up the most. When we focus on what God is eager to do, God stirs with great power. Let's look at what God has waited an eternity for.

Since there are some patterns of explaining that are more effective presentations of the Gospel, then God blesses some styles of preaching more than others. God does not always give the most effective styles of preaching to all. Likewise, since there are some patterns of explaining less effective salvation, then God blesses some styles less than others. Following are ten typical styles of presenting Salvation that dad avoided. Some styles are on the right track, they just stop short of what God is ready to do. Other styles are the closest imitations that falsely give the appearance of Salvation. The goal is to analyze preaching to identify what promotes rejection of Salvation, missing the mark altogether, being lukewarm, and a blazing fire. May God grant us the opportunity to come face to face with Him in all His Majesty.

Part of defining what God blesses is defining what He does not bless as much. The following are styles that dad avoided as the central theme in his search to give the Spirt room to work. These styles hold the key to being blessed more by God and passing along a greater blessing. Just trying to comprehend that the unsearchable is at our fingertips for God looks for excuses to lavish His love on us. May God reveal His ways and talk to us with greater power as we investigate what God does and does not bless.

Study of 10 Responses to Salvation

Response #1 – Time. When people die, they will end up in heaven or hell. Upon death, there is no chance to change the mind. Preachers claim, "You need to accept Christ now for you do not know how long you have to live!" The problem is that the healthy without a disaster looming very often has no sense of urgency. Too often the choice is, "Yes, I need to accept Christ someday before I die." Time is urgent for some; but for many, time is not urgent so they postpone the best decision they could ever make.

Dad lost some young strong and healthy that he loved dearly. Two of them are mentioned in the book: the caretaker and the one killed by a snake whom I got to take his place in the car. He would mention them in his sermons and make a strong appeal, but the possibility of dying was never the main focus of his plea for salvation.

19 I will tell my soul, "Soul, you have many goods laid up for many years. Take your ease, eat, drink, and be merry." 20 "But God said to him, 'You foolish one, tonight your soul is required of you. The things which you have prepared—whose will they be?'

-Luke 12:19, 20 (WEB)

Response #2 – Dreams. People have plans to be great someday. Everyone knows that they are in rebellion against God when they live for themselves. If all they are offered is Salvation, then surrendering to God is seen as losing their dreams. People often choose to postpone accepting Christ for a while so that they might live a little. What they do not see is that God has better plans, much better. Living their dreams means rejecting God's incredible wonders. Not following God's dreams leads to incredible soul poverty.

> *10 A little sleep, a little slumber, a little folding of the hands to sleep: 11 so your poverty will come as a robber, and your scarcity as an armed man.*
>
> *-Proverbs 6:10, 11 (WEB)*

Response #3 – Lordship. Lordship is a change of ownership – I surrender my will and let God be in control. If the hearer is offered a contract in which salvation is exchanged for accepting Christ, then the contract is complete upon accepting Christ. According to the contract, all someone must do is accept Christ to enter into heaven. Some sincerely say the prayer, expect heaven and will be greatly disappointed. If God is not made Lord, then the contract is not complete even for the sincerely repentant. Salvation has two elements, not one. Salvation is turning from living for self to living for God. If someone is not God's, how can God save the one who rejects being His?

If a woman says to you, "I want to marry your son because I want his money and fame, but I don't want any of the responsibilities of being his wife!" How would that go with you? Would you intentionally give your child to a gold-digger?

The Father's goal is to give His Son to the Church as a bride. A sincere prayer of salvation without lordship is not Salvation. Although many when they say a salvation prayer do surrender their wills to God and receive salvation. Deeply regretting failures is not enough for Salvation. Tragically, how many say a weeping prayer of repentance only to find it rejected by God as insufficient? A prayer of salvation must have repentance and lordship.

14 What good is it, my brothers, if a man says he has faith but has no works? Can faith save him?

- James 2:14 (WEB)

Response #4 – Fear. Some preachers are known as hellfire and damnation preachers. Their main motivation for salvation is negative. A little fear can go a long way in waking people up to reality. However, when the focus of the preaching is negative, some unique results appear. Fear is a very effective driving force for thrusting people away from danger, but fear is incompetent at steering. Some are very terrified of how bad they have been but are not ready to make Christ Lord. Fear is a great guide to avoiding some things, but fear does not lead down the right path.

When caught in a burning building, the scared do not care where they go. They just want out of the building. It does not matter if they reach a car, another building, or the street, they just want out.

When someone is scared of hell, fear will drive them away from the fire, but not necessarily drive people to God. When the motivation for saying the sinner's prayer is fear-driven, then God may not be embraced in the attempt to stay out of hell. Once again, many say empty salvation prayers. Saying a prayer of salvation to escape hell without surrendering to God is still useless for some fear is often a needed motivator; fear alone is not enough. People sometimes say a salvation prayer to avoid hell without the intention of making Christ Lord.

19 You believe that God is one. You do well. The demons also believe, and shudder.

- James 2:19 (WEB)

Response #5 – Guilt trip. Guilt trip preaching focused on needing to do better. The object is neither to get to heaven nor to avoid hell, but to be great. The ones driven to be great are very vulnerable to guilt trip evangelism. Those who work at being great are deafened by the roaring of guilt. They are hopelessly caught between, "I've got to!" and "I can't!" Guilt trip evangelism is exciting news to some for they see salvation as the most perfect opportunity. The understanding is that God will erase all sins, and life starts over. Some pray a salvation prayer like this one, "God forgive my sins so that I can do better." Guilt trip salvation is driven by two overwhelming needs: to erase past failures and to prove greatness. Guilt-ridden glory seekers are not looking for personal relationships; they are looking for fans. Fame seekers are always saying something like this, "Watch, look what I can do!" " Look, I can do it!" "That is not fair! That one does not count!" "I have something to prove!" "Don't you dare get in my way!" "That makes me so mad!" "Wait, I'll do better next time!"

The glory seekers are always splashing things like bitterness, frustration, taunting, bragging, gloating, and defensiveness. Those who try hard can seem like Godly people until someone gets in their way of proving greatness. Guilt is God's way of saying that things are messed up and there is a need for salvation. Guilt trip evangelism can dangerously lead people only close enough to God to seek a do-over. Some get so close yet remain so far from salvation. To this point, there are several reasons why people will say repentant prayers that are ineffective for salvation: to avoid hell, get to heaven, be rid of guilt, and get a chance to start over.

21 "Not everyone who says to me, 'Lord, Lord,'
will enter into the Kingdom of Heaven, but he who
does the will of my Father who is in heaven.

-Matthew 7:21 (WEB)

Response #6 – Obligation. Many people combine Christianity and obligation. The two do not match. Christianity and obligation clash. How many times have we all heard sermons where we are compelled to tell others about Christ? It is our duty as Christians to tell others about Christ, so we hear. Once living for God becomes an obligation, faith is lost. Obligations evict faith for self. Obligation and faith do not co-exist. Obligation means, "I have to do it!" Faith means, "I trust God to do it through me!" The obligation is the Pharisee's imitation of faith. The self-righteous are obligated to carry the burden of the faith.

If something is truly a godly task, then it takes a miracle to do it. If a miracle is not needed to accomplish the task, then only human strength is needed to accomplish this human task. To do a Godly act, then God must intervene with a miracle for the humans to behave godly. Witnessing is a godly act that can only be partially imitated by men. Witnessing can happen without God's help to a point. Humans can say some eternal truths through human strength. No one can say what God intends without a miracle. Being led to say the right thing is a miracle. Without God's leadership, people are blind to the calling. No one understands the mind of God unless God speaks. Staying in the calling requires constant miracles. Human strength is hard work for obligations are wearisome in a hurry. Motivation runs out long before God's task is finished. For the obligated to keep going, excitement is replaced with pushing hard. Pushing hard leads to exhaustion. Exhaustion ends in burnout. When burnout happens, there is no strength to finish the call as seen through human eyes. Obligation seems noble but it is a selfish imitation of faith. Obligations are attempts to do Godly things in human strength for personal glory. Obligations never focus on God's plan or His words for the moment.

Obligation and the drive for greatness are terribly similar. Both have something to prove, but the motivations are different. An obligation is driven by a sense of responsibility while the search for greatness is proving merit. Teaching obligation is teaching to rely on self and not on God.

18 For I know that in me, that is, in my flesh, dwells no good thing. Desire is present with me, but I don't find it doing that which is good. 19 For the good which I desire, I don't do; but the evil which I don't desire, that I practice. 20 But if what I don't desire, that I do, it is no more I that do it, but sin which dwells in me.

-Romans 7:18-20 (WEB)

Response #7 – Fitting in. It seems logical that fitting in is a good thing. However, fitting in caused mom and dad more grief than anything else. Fitting in is a paradox. Fitting in is trying to find a place to belong. Belonging sounds good, but it is a terrible end goal. Fitting in or belonging is all about me. Fitting-in answers the question, "How can I make my life secure and satisfying?"

One of Jesus' favorite themes was that to save your life, you must give up your life. Those who choose to keep their lives will lose them. The one, who gives up life into God's hands, chooses to not be concerned about fitting in on earth. Fitting in is a very strong need, but real belonging is fitting-in with God.

Those who play the fitting-in try to be in the middle with others fitting in around them. Fitting-in divides people into four groups: core, edge, ignored, and persecuted. The core is a small handful that keeps each other in the middle, and the rest further out. Fitting in is establishing a pecking order where the self and the inner group are at the top. The rest of the people are assigned rankings establishing how close to the core they belong. Most are given enough value so that they will stick around, but not enough value to feel like they fit in. Those who seek to fit-in rank people's values as high, medium, low, no, and minus. The ignored neither offer value nor are seen as threatening to the in-group. The crushed can offer a lot more threat than benefit to the in-group. Oftentimes, the crushed offer their best to the in-group, but crushing people gives the powerful something to bond over. The fitting-in game ranks people by value to the group.

It could be said that fitting in is not an issue in the presentation of the Gospel for salvation. The need to fit-in causes two great problems in witnessing. One problem with fitting in is that the most responsive people to the Gospel are usually the outcasts. Fitting in causes those who would be most receptive to the Gospel not to hear the Gospel. If the outcasts did hear and respond, they soon find out that they are still outcasts. A second problem with fitting-in evangelism is the reach to those who pass on the same values forever pushing the hurting masses away. Fitting-in people reach out to those who are good at the fitting-in game. Fitting-in evangelism safeguards selfishness at the expense of those who Christ died for. Those who try to save their lives by fitting in lose it.

11 But he who is greatest among you will be your servant.

-Matthew 23:11 (WEB)

Response #8 – Notching the gun. In the old Wild West, it is said that some would cut out a notch in their gun every time they killed someone. The notches are bragging rights of cruelty. Some who are particularly good at evangelism try to notch their gun. They know how to present the Gospel in such a way that people respond. They keep a tally proving their greatness. The downside is that love is replaced with cruelty. Those who are hearing the Gospel are being convicted by God. The Gospel presenter is far more interested in bragging rights than saving the lost. The sad part is that the lost hear both the wonderful Gospel and the arrogant stench. Many respond to the Gospel despite the presenter. The day always comes when salvation notching is deeply regretted. God will bring every evil to a day of accounting.

> 22 Many will tell me in that day, 'Lord, Lord, didn't we prophesy in your name, in your name cast out demons, and in your name do many mighty works?' 23 Then I will tell them, 'I never knew you. Depart from me, you who work iniquity.'
>
> -Matthew 7:22, 23 (WEB)

Response #9 – Discipleship. A Lot of people say a sincere salvation prayer without walking nearly as close to God as they intended. Imagine a church where no one has a vision of what God wants to do. Someone accepts Christ as Savior, and everyone stays in their little comfort zone. At first, the new Christian tries a few times to reach out but does not have the skills or the opportunities to be very successful. After a while, the new believer realizes that everyone else seems to love the Lord, but they are disengaged from a vision of what God can do. The unfocused church is training the new converts on living an unengaged faith. Real discipleship is becoming like Jesus in words and actions. Real discipleship is a planned approach to developing ministry skills and an understanding to match. Jesus spent much time drawing people to the Father outside the walls of the church.

Imagine an American football game where the team huddles around the coach with all the pads, and helmets, then the team refuses to take the field. The players show loyalty and greatness by attending the huddle and staying on the squad. These players diligently study the playbook, but there is always an excuse for why none of them can take the field. The typical excuses can be like these: an injury, do not want to aggravate an old injury, being too slow, tired of losing, or don't want to get sweaty because there are plans after the game.

Many churches huddle around God by going to church. They study the playbook by reading the Bible. Unless someone attends church and sees them, no one knows that these people are Christians. When the plan of Salvation is presented without explaining discipleship, then the new believer often assumes that discipleship is irrelevant and that salvation is all that matters.

26 For as the body apart from the spirit is dead, even so faith apart from works is dead.

- James 2:26 (WEB)

Response #10 – Semantics. Do you know what semantics means? If you do not, you just made my point. Semantics refers to the meaning of words or phrases. I could have simply said, "meaning of words," but then I would have been wrong even if the meaning is right. This is a big deal when preaching. Using the right words, excuse me, using the proper words as some would correct me. Others would say, "Now you are sounding all educated, and you don't talk like common folk." Some may have the hardest time with this point. Actually, this point is two points, but I am going around in circles just like some of these arguments go. Many have been extensively taught the proper way to present the plan of salvation. The teaching is that certain phrases need to be said when presenting the Gospel. Many believe that without these words, it is just about impossible to be saved. Some of these keywords are "sinner," "sinned," "sin," "salvation," "born again," "accepted Christ," "accept Christ into your heart," "receive Jesus into your heart," "wages of sin are death," "redeemed," "for by grace you are saved," and there are many more. These very words are so crystal clear yet equally ambiguous.

Some people indeed start trying to make sense of things by finding the most important terms and defining them. This is a very common way of thinking. Huge meaning is attached to these key phrases when these words are spoken much thought and emotion are stirred.

The first semantic question is how important are the words for salvation? For someone who grew up in the church and heard these words often, these words may have a lot of meaning. For someone who has not had any exposure to Christianity, the words can make salvation confusing. When these phrases are all jumbled together, then the meaning can easily become fuzzy. Many would argue that the phrases would have to be taught so Salvation can happen. Many argue that unless the words "I accept Jesus into my heart" is said, salvation cannot happen. The challenge is to find the phrase "I accept Jesus in my heart" in the Bible. If that phrase is that important, why did God never use it in the Bible? For those who need the proper words, Jesus' teachings are particularly troubling. Jesus did miracles, and people believed. The Bible does not record Jesus leading people in Salvation prayers. The listeners responded in faith to healings, parables, and the teachings of Jesus. Multitudes were heaven-bound without a formal Salvation prayer or salvation phrases.

Missions have some unique challenges. People want to hear the salvation Gospel in words that are most comfortable to them. People want talk that is the most like their own. Now add in the challenge of speaking different languages. Languages do not match up perfectly. The terms that are so important in English may neither have a good equivalent in another language nor the same cultural understanding. If there is not a good equivalent for this must-have Salvation phrase, does that mean that they cannot get saved using another language? When a Salvation prayer is said without the "right words," many long-time Christians are deeply concerned that the prayer is not the real deal. It is a problem if the newly saved are seen as not saved by veteran Christians. Semantic problem #1 is "What words or phrases should be used?"

There is a semantic problem 2. These key phrases do not mean the same thing to all church members. In churches, there is an 80/20 rule. On average, 80% of the work done, and the money given, is by 20% in a congregation. I am not saying how valid the 80/20 rule is, but that it makes an incredible observation. The point is not to argue ability and wealth, but desires. It is sad when only about 20% contribute their efforts to the church while the other 80% enjoy the labor of the 20%. To sum up the 80/20 rule, 80% think about how the church can serve them while 20% think of how they can serve the church. Once again, these are only perceived averages and could vary significantly.

It appears that the subject jumped from semantics to the 80/20 rule. Semantic point 2 has been made. People often see service and giving the same way they see salvation. The understanding of Salvation phrases reflects the outlook on the role of the Christians in the church. Some see Christianity as a calling to give life away, while others see Christianity as being served, and these salvation phrases are no different. Some see "accepting Jesus" as giving their lives over to Jesus, while others see it as gaining something from God. Some are saying, "I am accepting Jesus as Lord of my life" while others say, "I am graciously permitting God to come into my life on my terms." In Christian churches, the fake often lives side by side hidden by the official slogans.

The duo meaning of Christian terminology makes both the seekers and pretenders comfortable together. Christianity should not lose the words that have proven so significant, but there are problems. Neither words save nor do they have a consistent meaning. The semantic issues are of importance since the words/phrases have contradictory meanings and hide a multitude of sins.

20 Seeing their faith, he said to him, "Man, your sins are forgiven you."

-Luke 5:20 (WEB)

Dad's Preaching Advice 2 of 3

Like before, dad's preaching advice was as simple and short as before. He said it and he was done. It made so much sense that there was not much discussion. Dad said, "When preaching, the bigger the carrot you dangle in front of them, the more likely they are to respond." Done! This was a new and fascinating concept to me, but I saw that dad effectively used it. I repeated back, "The more you offer them, the more likely they are to respond." When all that is presented is Salvation, no carrot is presented.

19 "Repent therefore, and turn again, that your sins may be blotted out, so that there may come times of refreshing from the presence of the Lord,

-Acts 3:19 (WEB)

Dad's Preaching Advice 3 of 3

Dad's third advice as you have probably figured out by now was as short as the others. Dad said, "Don't beat people over the head too much, for you will make them bitter." My dad was right. Constantly fighting sin only increases negativity. I asked dad, "So what do you do if they continue in their sin?" Dad answered, "Fighting them only entrenches them in their sin." Fighting sin is not the answer.

My dad has confronted many people about what they are doing wrong. Fear and guilt need to play their part. Dad was always for the person and worked at finding better solutions. Dad knew that the object was ending sin and drawing people closer to God. I was constantly amazed at how Dad could refocus people.

11 Does a spring send out from the same opening fresh and bitter water?

- James 3:11 (WEB)

JIM H DARNELL JR

Chapter 8

24 Hours

God, Dad, and the Carrot

Dad often spent most of His message talking about God's wonderful plan for their lives. It was fascinating to watch dad glow with excitement as he revealed the wonders of the calling. Dad would carry along his listeners. Dad had already shown great compassion for them as he made his rounds greeting people. Dad was already believable. When he told these searching people the answer to their dreams, they were eager. The crowd usually got very excited. They had a chance to exchange their aimless wanderings from the village to the city for a bigger-than-life calling.

The listeners went from not understanding God's vision to understanding it, to wanting it, to eagerly desiring God's wonderful plan before the sermon ended. They were ready to live for God even before they were saved. The reward was huge, the people were ready to jump at the Gospel.

All my dad spoke was words, while the Holy Spirit directly plugged into their emotions and thoughts. Something far greater than dad was happening for God was gripping hearts often with words like these. When dad said, "God has a plan for your life," God said, "I sure do!" When dad said, "God wants to reach through you as far as God can reach," God said, "Try me and see how far I can reach!" When dad said, "God wants you to be His hands and feet," God said, "I want you to be the one!" When dad said, "God has set aside people for you only to reach," God showed images in their heads of lost and hurting loved ones. When dad said, "God has called you to be His ambassador to the world," God spoke, "Be my ambassador!" When dad said, "You and God make a majority," God said, "Who can get in our way? We dominate!" When dad said, "God's plan is the most wonderful thing in all the world," God spoke such incredible wonders to the heart that no words can describe. When dad said, "God will be with you always. He will never leave you," God declared, "I will be with you always." When Dad said, "God loves you," God overwhelmingly spoke of love. When dad said, "From this day forward, you will be on a mission for God," God announced, "I have a mission for you every moment of every day!" When dad said, "God has chosen you to reach Ivory Coast for Christ," God said, "I want you to reach Ivory Coast in my name." When dad said, "God will show you what to say," God echoed, "I have words ready for you to speak!" When dad said, "When people reject you, God will be there," God said, "Rejection is coming, but I will show you more acceptance, far more!"

God speaks however He wants. However, God regularly drives home the points that interest Him. God took the carrot that dad presented and highly touched many lives. What dad did was offer the very things that God blessed dad with most.

28 We know that all things work together for good for those who love God, for those who are called according to his purpose.

—Romans 8:28 (WEB)

God, Dad, and the Gospel of Salvation

At this point in dad's typical messages, many were convinced to become all that God could make them. They were excited about God's plan. God had been screaming within them, "Yes!" The call to reach their loved ones and the country for Christ reached home even before salvation occurred. Dad had not yet explained Salvation.

Salvation was presented as an entry into a wonderful life. Dad would say something like this, "If you want God to reach through you as far as God can reach, then you must invite God to come into your life." Salvation was presented as the entry into the most wonderful life with God. Salvation must happen before anyone can receive God's love and become His ambassador.

20 Behold, I stand at the door and knock. If anyone hears my voice and opens the door, then I will come into him, and will dine with him, and he with me. 21 He who overcomes, I will give to him to sit down with me on my throne, as I also overcame, and sat down with my Father on his throne.

-Revelations 3:20, 21 (WEB)

Readiness to Accept the Gospel

So many Africans were ready to accept the Gospel right then. They did not want to wait until later. They were sold on the idea that they finally found what they were looking for. It was not the fame and riches of the city. They longed for God's love and His purpose for their lives. Nothing else mattered. You could tell many of them were ready right then to make their choice without wanting to wait until the end of the sermon. They had nothing to gain and only to lose if they did not become God's right then.

17 For God didn't send his Son into the world to judge the world, but that the world should be saved through him.

- John 3:17 (WEB)

Family Value

Don't you like it when I get to good parts and then change the topic? Here is one of those times. I remember my best friend Kojo (Koadjio pronunciation). We were talking one day, and he said something strange, "Without family, you are nothing!" I had a hard time understanding what he meant. I had given up my family except for my mom, dad, and sisters: Patty and Julia. I had chosen to leave behind so much, but I did not feel as if I was nothing because I had not lost all my family. I would have felt like nothing if I had lost my parents and sisters. I would have been so alone; I don't know how I would have lived.

In Africa, everything revolves around family. The Africans have a lot of holidays, and traditions. They depend a lot more on each other. They live in the family's village. Their protection from others is in family numbers. The jungle has countless vicious animals from insects to hippos that the village helps control and avoid. They tell stories of their dead relatives and the greatness of their deeds. These deeds define their social standing among other families. Africans strive to live up to their heritage and pass it on.

I saw myself as the son of missionaries who were on a mission for God. I experienced the wonders of God's love. If my identity were taken away from me, I would have felt very lost.

As the years rolled on, most do not know what mom and dad have done. Even if I tell people, they rarely can identify with my childhood. I have never once been asked for my point of view on how to solve a church problem because of my unique heritage. I now live in a small enough community where those who grew up in this community often get preferred treatment. I now understand what Kojo was saying. If my parents had the same kind of success locally as they had in Africa, I would be highly esteemed and have many more opportunities. God chose for me to experience both: what it was like to have a wonderful heritage and not to have a heritage.

Africans have traditionally been very spiritual people. They worship the same way that their ancestors had worshiped. Because they worshiped like the ones who went on before them, they honored their ancestors. Every celebration and action in life had a spiritual meaning. Worshipping like their ancestors meant accepting their heritage.

There is an additional element to what Kojo was telling me. Survival often requires family ties. The few jobs that existed went to family. Africans were often day laborers. When they worked, what they earned went toward buying food for the extended family. When they did not work, someone else bought the food. Life was too unstable for many Africans to survive without the help of family.

> 11 Not that I speak because of lack, for I have learned in whatever state I am, to be content in it. 12 I know how to be humbled, and I also know how to abound. In everything and in all things I have learned the secret of both to be filled and to be hungry, both to abound and to be in need.
>
> -Philippians 4:11, 12 (WEB)

24 hours

Okay, we are back on track like we were never off track. Twenty-four hours was a real game-changer. Dad went on to explain something that shocked and terrified many. Dad said, "You have 24 hours to tell your family about your decision to accept Christ." This clause is not a part of salvation, but this clause would definitely weed out many who were not willing to live for Christ. Faith in God is proved by actions.

I remember one Sunday morning when dad was preaching. I saw something like I had never seen before or since. What I saw forever impacted my life. I was staring at one man. At the time, I did not understand why I was fascinated by this man that was slightly behind me, but I was in a perfect position to stare at his face. Dad gave his usual clause, "You have 24 hours to tell your family about your decision to accept Christ." This man's face looked like it had been slapped. The wrinkles of a hand impressed upon his face. It was like I was looking through an invisible hand as it left its imprint upon the face. From a child's perspective, the imagery was cool, yet terrifying at the same time. This man knew that if he told his family about a decision to follow Christ, then he would likely be killed or disowned. Terror invaded this man. One way or another, this man's life was over the way he knew it. If he rejected God, then he would live the rest of his days in the shadow of his coming doom. If he told his family, then he would become nothing.

God in His wisdom knew that I should not know this particular man's fate. I do not know if he accepted Christ that day or not. I do not know if he walked away from the most wonderful discovery of his life. This man was one moment in time representing the choice that countless Africans have had to make.

I cannot tell you more about the man, but I can tell you about the end of that particular service. I had the opportunity to see the most beautiful sight. Hold on, we are not there yet.

33 But whoever denies me before men, I will also deny him before my Father who is in heaven.

-Matthew 10:33 (WEB)

Presentation of Salvation

Dad spelled out how one must become saved. The Africans had no concept of who Jesus was or of the Salvation that God offered.

Many West African tribes had the concept that there was God above all the spirits, but there was a problem. Their beliefs provided for absolutely no way to reach God directly. God could only be reached through the spirits.

Dad simply explained the Gospel. Dad said, "God wants a friendship with you." God echoed, "Oh I most certainly want friendship with you." God has a way of making people feel the warmest and coziest ever on the inside. Dad said, "Your wrongs have made God so mad that He cannot accept you the way you are." Can you imagine the flood of guilt and failure that rushed over at that moment as God flashed the most shameful and embarrassing moments of life? Dad continued, "God became so mad at you that it would cost you your life forever in hell." God declares, "That's right! You deserve hell." It is a wonderful thing that the story does not end there. Dad explained, "A price had to be paid for your faults before God can accept you." God agreed, "A price has to be paid before I will take you as my own." Dad continued, "God sent His Son to die in your place." God declared, "My Son died for you so that you do not have to die in hell." Dad moved on, "To become saved, you have to accept that Christ has died in your place." God confirmed, "Jesus died for you!" Dad said, "God wants to enter your life and live within you." God reaffirmed, "Yes, I want to live within you forever." Dad clarified, "You will become God's child." God assured, "Yes, it is just like you were born in my family." Dad declared, "You will gain the full privileges of being His." God promised, "Yes, you will become fully mine." Dad continued, "To invite God into your heart, you must ask God to forgive you of your sins." God affirmed, "Yes, you must ask me to forgive you." Dad went on, "You must invite God into your heart and live with you forever." God echoed, "Yes, for me to come in, you must invite me in." Dad warned, "God will not come in unless you invite Him." God agreed, "I am waiting; what is your choice?" Dad explained, "When you invite God into your life, you are inviting Him in to be the Master of your life." God made clear, "I will come if you let me be in charge." Dad bluntly explained, "God wants every part of your life to be His." God simply said, "I want all of your life." Dad declared, "Once you invite God in, you can begin your journey with Him." God confirmed, "I have a wonderful plan for your life."

Salvation is a transfer of ownership. The horrors of rebellion have to be realized. Rejecting God to follow personal dreams angered Him. The Almighty had a plan for all, and He was told, "No!" God's wrath must be settled. Jesus' payment for sin must be received so that eternity is not spent in hell. God so desired to call people unto Himself that He sent His own Son Jesus to pay our price. The journey with God begins with repentance. Sins must be confessed. God wants to be invited in. Part of salvation is not just repentance but also making Him Master. God seeks full surrender to Him. Salvation is turning from wrong and making Him Lord of all.

> 9 By this God's love was revealed in us, that God has sent his one and only Son into the world that we might live through him. 10 In this is love, not that we loved God, but that he loved us, and sent his Son as the atoning sacrifice for our sins.
>
> -1 John 4:9, 10 (WEB)

The Invitation

had expressed how much God loves them. A vision had been laid before them that they could be the ones who bring salvation to their family and the country. They were told that before they could become God's, they had to trust Him to be their Lord and Savior. They were given a 24-hour challenge to tell their family about their decision. The truth was set before them.

Dad would often finish his invitations with statements like this one, "Will you commit before God that you will follow Him no matter where He leads you or takes you? As God would fill them with wonder, He would ask, "Will you go and do whatever I ask?" Dad declared, "Today is the day to accept Christ." God passionately expressed, "Today is the day!" Dad would say, "Today, you may leave this place and never hear the Gospel presented again." God says, "You may never hear my offer again."

2 By this you know the Spirit of God: every spirit who confesses that Jesus Christ has come in the flesh is of God, 3 and every spirit who doesn't confess that Jesus Christ has come in the flesh is not of God, and this is the spirit of the Antichrist, of whom you have heard that it comes. Now it is in the world already.

-1 John 4:2, 3 (WEB)

Now Is the Time and a Deadly Snake Bite

There was a story that made such a huge impact on dad that he told it many times. Dad had gone up to a village and presented the Gospel. Eight people had accepted Christ on dad's visit. The village was far away, and during the school year, I could not go. It was vacation time, and I pleaded with dad to take me on this follow-up visit. Dad was going back up to baptize the believers, but the car only seated nine, and there would be no room for me after dad picked up the believers awaiting baptism. The villagers briefly called Dad explaining that one had died. Dad and I rode in the car all day Saturday and spent the night.

When dad got there Sunday, the villagers told the story that two days before, one of the new believers was bitten by a snake and died. Amid their deep sorrow, the villagers were excited because the man had made his choice. Now he had entered eternity with his Lord. Strange isn't it, this obscure man who spent only a few days with the Lord on this earth, made an impact upon countless people. The man surrendered to God and chose to risk it all. He knew he was likely to die or greatly suffer at the hands of his family, yet he made His choice. God simply told the man, "You have proved you have risked it all for Me. Come home to your great reward!" The mix of God's great pleasure and urgency of the Gospel makes this man's story so powerful. This barely on-time rescue by the Almighty was such a motivator for dad. This story so gripped him that he told it over and over again. "Now is the time!" has rung home for countless listeners. I do not know the impact this book will have, but God has chosen to spread His pleasure of this man's choice. This man's death had a huge impact on me. When the men rode to become baptized, there would not have been room for me. I sat in the dead man's seat.

The untold part is so significant that it cannot be ignored. There is more to this story than a man willing to lose it all and dying.

There is the part about the dying process. The Africans have watched others die of snake bites; others are terrified because they had not made peace with God. How dreadful it would be to know that only a few moments are left in life only to die in eternal fear? Facing an angry God is a terrifying way to die. With the full weight of God's guilt pressing down on the unrepentant, "You are going to hell because you are rejecting me!"

The villagers kept repeating over and over, "One of us (Christians) is no longer with us. He was snake bitten and died." At the time, they were not ready to tell more for they were in shock. The small band of Christians wanted so much for their family to find the same Wondrous God.

At first, in African culture, it could have been said that he cursed himself by rejecting the spirits. His death told a different story. In all the suffering, God still delighted the man He was welcoming home. His death was to shine God's glory. Rather than fading away in a cursed manner, this man died touching the glory to come. Indirectly, the people saw a glimpse of the Almighty. Salvation had come, and Salvation was thumping on every heart.

The 24-Hour Consequences

Imagine for a moment that the 24-hour challenge was not accepted. Christians would have hidden their faith like it was a curse. They would have lived like they were covering up a terrible scandal. Imagine the snake-bitten man who has refused to tell his family of his decision. Once the snake bit, what could he say to demonstrate that Christianity was honorable if he had hidden Christianity like a dishonor?

On the contrary, it was because God was so precious to them that the 24-hour challenge was a must. The rejection, persecution, and even martyrdom, God displayed His Glory, and many believed. Once the decision was made to honor God with the 24 hours, the choice was made to sacrifice all for the Eternal. God displayed His great pleasure in the faithful. The desperate people have found far more than they could ever have imagined in trusting their Maker. God's glory shone in and through the faithful. The new converts went from the depths of despair to heavenly heights in the snap of their decision.

What the original believers saw in mom and dad, was contagious. Others walked the incredible journey with the same infectious joy. 24 hours made all the difference.

4 But many of those who heard the word believed, and the number of the men came to be about five thousand.

-Acts 4:4   (WEB)

Chapter 9

The Invitation

The Invitation Prayer

Dad would tell them in advance the prayer that they would pray if they chose to give their hearts to God. The prayer went something like this:

Dear God, I am sorry for my sins. Please forgive me for the wrongs that I have done. I accept Jesus' payment for my sins. Please come into my life and be my master forever. I choose to follow You wherever you lead me. Thank you for forgiving me for my rebellions. Thank you for sending Jesus to die in my place. Thank you for coming into my life and for having called me to be your hands and feet, in Jesus' name, amen.

Dad clearly told them the prayer that they would pray so that there would not be any surprises.

9 If we confess our sins, he is faithful and righteous to forgive us the sins, and to cleanse us from all unrighteousness.

-1 John 1:9 (WEB)

The Invitation

Dad would finish with a strong challenge which often went like this.

"In a moment I will invite you to come forward to make God the Master of your life. He has a wonderful plan for your life. Are you willing to come and make your decision today? Do not hesitate for the time is now. Are you so willing to give your all to Him that you will be willing to come up immediately when the music starts? Are you going to follow the example of others who have chosen to give their all to Christ or are you going to set the example and come immediately when the music begins? Do not hesitate. Come. God is calling you. Come. Come as God calls."

15 Whoever confesses that Jesus is the Son of God, God remains in him, and he in God.

-1 John 4:15 (WEB)

The Invitation Response

These responses were my favorite time in Africa. All the hardships of leaving family behind, and the personal cost of living in Africa were worth it for the times of invitation.

I told you that I had a story to finish about the invitation. As I said, I do not know what happened to the young man whose face contorted like he had been slapped. There was a reason why I did not know. When dad said, "Come!" The people came. All over the building people hopped up. Many tried to be first. They were all firsts because they rose at the same time. Some arrived at the front before others, but so many chose to come.

Many were so eager to come that they would lunge forward, stop and back up only to do it again. They were so ready to lose it all because the gain was so much greater.

All the false starts so stirred me. Others just needed to see the brave ones stand first. Still, they were all brave, they knew about tomorrow. They stood. They made their way to aisles and walked forward. One after another chose to surrender and make their way forward.

After a while, there was a congestion problem. The front was not big enough. The front of the church disappeared from view because so many came. I was so fascinated by the many people who were risking everything that I lost track of the man behind me. Dad left the invitation open as long as people were coming.

How many people came that day, I don't really know, but mass decisions were the norm. Large numbers usually committed their lives to God every time day preached.

Miracles had formed within the hearers. They listened with great intensity. Many were captivated by the incredible visions of what God sought to do. They were spell-bound. They believed that they could do their part in winning their families and country to Christ. They made their stand for God.

> *32 The multitude of those who believed were of one heart and soul. Not one of them claimed that anything of the things which he possessed was his own, but they had all things in common.*
>
> *-Acts 4:32 (WEB)*

Living the 24 Hours

I just can't imagine the heartbreak of the 24 hours. Having God on their side did not make their pains disappear. The agony of the 24-hour journey and beyond was immense.

God healed many pains, but they experienced the pains of carrying their cross. When God said, "Who are your favorites – your family or me?" God made the Africans choose. God brought tension between the new believers and their families. God added incredible wonders and much sorrow of rejection and helplessness.

17 In this, love has been made perfect among us, that we may have boldness in the day of judgment, because as he is, even so, we are in this world.

-1 John 4:17 (WEB)

Chapter 10

The 24-Hour Cost

Joseph

Joseph was the African man I knew best. He always loved us kids and had time to play with us. He was a Muslim when we first met. Joseph was one of the very first who came to Christ in mom and dad's ministry.

His salvation came at a great price – renounce Christianity or die. Forty men had pledged that they would not eat until they had killed Joseph. I figured that Joseph would be crushed by the crisis and would be all mopey. Every time I saw him, Joseph had a huge belly laugh. He hopped around like he had the energy to burn rather than carrying the death sentence. Joseph was in great sorrow and great joy at the same time. Joseph was eager to live for His newfound God and willing to die for him as well. The only time that Joseph went outside was when he was surrounded by a large group of believers.

One day, my dad had enough; he sought out those men. Dad went into an African neighborhood all by himself to confront this band of men. Dad told them, "We know who you are. If you kill Joseph, the police will find you." The men thought about what dad had said and realized that they could not kill Joseph and get away with it. They responded, "Okay, Joseph can live. We will no longer try to kill him."

From that day on, Joseph traveled freely by himself. Not one thing changed. Joseph was full of joy after the crisis; and, yes, he was overjoyed by the mercies of God during the threats on his life. Joseph somehow kept a smile on his face and lived with God splashing all sorts of love on him.

For Joseph, the hard aches were far from over. His family rejected him and cut him off as well. His old friends had tried to kill him. He was nothing according to African culture for his family and friends treated him as if he were dead.

About five years later, he traveled with us nearly halfway across the huge African continent to try to convince his parents one last time to make peace. Upon Joseph's arrival at the village, his dad said, "It is so good to see you, Joseph!" Joseph answered, "It is good to see you, dad!" His dad asked, "Are you coming back to the faith of your fathers?" Joseph answered, "No!" His dad asked another question, "Have you come to bring me some money?" Joseph said, "No dad, I just came to tell you that I love you." Joseph's dad continued, "Have you come to give me a car?" Joseph answered, "No dad, I do not have a car to give you. I just came to tell you that I love you." His dad said, "You have not come to make peace with your fathers. You have not come to give me money or a car. What have you come for?" Joseph earnestly pleaded so that his dad would understand, "I have come to tell you that I love you." His dad raised a bony finger and pointed at the gate, "If you did not come to make peace with your fathers, give me money or give me a car, then get out!" The conversation was over that quickly. Joseph left broken-hearted. From the time Joseph left us to the time, he came back, it was about 15 minutes. The time also included walking time to and from the village.

Joseph's dad would have taken a major gift as an entrance fee back into the family. Joseph did not want to buy his way back into the family. Joseph wanted to restore the relationship with his family and see them make peace with the God that Joseph found worth living for.

On the long journey backto Ivory Coast, Joseph just stared. He did not joke with us kids according to his habits. I did not see it, but dad watched Joseph cry. Dad only saw two African men cry while we were there.

Joseph had learned what it was like the previous five years to be a nothing in African culture. He had hoped for a change; instead, he learned a whole new level of what nothing is.

I only remember seeing Joseph once after that. It was several years later. Dad had driven up to Joseph's house. Joseph was out front just having the best of time when we arrived. Dad asked me to stay in the car. I did not talk to Joseph which made me sad. I wanted to joke with him one more time. Even then I realized that it didn't matter. Joseph was happy.

21 "Brother will deliver up brother to death, and the father his child. Children will rise up against parents and cause them to be put to death.

-Matthew 10:21 (WEB)

A Little Bit More About Joseph

Joseph is the one holding me on the front cover. Joseph made me feel so treasured. He was there with us at many baptisms. He participated in all the new church plants at the beginning. He was new to the faith, but he chose to go around with mom and dad to start churches. Tens of thousands have come to Christ through Joseph's church planting and ministry. As of 2019, he was in his 90s and still bounced around. He was pastoring one of the churches that we all planted together. His dad rejected him to the end; but his mother, when she was old believed in Christ. Joseph is my hero.

24 Hours Caused Rejection

Joseph experienced the two most common responses that

families gave the new believers. Joseph was threatened with death, and he was cast out of the family. I don't know how many died for telling their families about their love for Jesus. Mom and dad never talked about it; I wanted to ask, but I could not do it. I imagine that several did. The most common response was total rejection. They were told how unfit they were to be a member of the family because they rejected the faith and customs of their fathers. The new believer had to endure the words declaring total failure by their families. After the humiliation, the believers were cast out of their families. If they had a job related to the family business, they lost it. They also lost their lodging. These were incredibly sad and troubling days.

22 You will be hated by all men for my name's sake, but he who endures to the end will be saved.

-Matthew 10:22 (WEB)

24 Hours – What Next?

Many had nowhere to go after being evicted. No one in the family would take them in. They would return to the church. One of the church members would open his house up to the new Christian. There was not much to open up as most African houses were equivalent to small rooms. Here is what dad once said, "The believer would open up his home to the new believer. There were already 17 sleeping on the floor. They would all pick up their mats and move them around to make room for one more."

This is the tropics. The wind just doesn't exist, especially at night. The overhead fan (if they had one) would blow, but no wind could be felt. The temperature at night dropped maybe 3 degrees or down to about 90F on a cool night. The humidity was also between 90-100%. The heat of that many bodies lit up the room.

Can you imagine the bonds that those people experienced? They had all been kicked out of their families. No one had anywhere else to go. They were all stuck together, but that is only a part of the story.

They all experienced God's incredible love. No one there was a lukewarm Christian. They all had paid a price for which there will be 100 times reward in heaven, but it was God's thrill on this earth that kept them there.

40 "He who receives you receives me, and he who receives me receives him who sent me.

-Matthew 10:40 (WEB)

JIM H DARNELL JR

Chapter 11

An Abundance of Joy

Building an Auditorium

Dad was a robust man. He never lost the strength of his youth while in Africa. He might have lost some, but he still had more than enough. We went one Saturday to join a church that was building a place to worship. The whole church had gathered to join our family and build the building.

Dad, in his usual way, was working harder than any of the others there. They saw how strong dad was, so they decided on a contest. Their best against dad. That was not fair, we only had one man to choose from, and dad did not do manual labor for a living. When they worked, they all did manual labor. They picked their strongest man. He was young, in shape, and rippling with muscles. Dad was middle age or a little past and accompanied by too much gut.

They both loaded their wheelbarrows full of sand. They counted shovel scoops. Everyone was satisfied that both wheelbarrows weighed the same. There were two trails side by side from the sand pile to the worksite. This place was not like American worksites where a bulldozer came first. The ground was uneven and rough. Many of the low spots were filled with water. Under and around the water was mud. They had placed boards over the mud so that the wheelbarrows would not bog down. To make things worse, the trails were not straight. The trails followed the high points of land, and when there were not any high points there was aboard.

Walking a wheelbarrow over such ground is not as simple as keeping the wheel on the high point. Turning a wheelbarrow sharply meant that the wheel had to come to a complete stop. The person had to move to the side until the wheelbarrow was pointed in the right direction before continuing. Still, even that explanation is too simple. Moving to the side meant moving off the high points of land. Sometimes puddles had to be hopped over while holding up one end of the wheelbarrow. Other times, puddles had to be jumped with one leg ending in a straddled position. The boards were not that easy to navigate either.

If they went down to a board, then the wheelbarrow full of sand picked up speed. With too much speed, it was easy to run a wheelbarrow off the board and into the mud. The wheel would burry itself. If the wheelbarrow could be dragged, then the muddy wheel would be very uneven and difficult to push.

Some boards were laying on top of the ridge. A running start was needed to bounce the wheel over the board with such a load. Once again, such speeds made it quite easy to run the wheel off the skinny boards. Nearly every time they came to a board, they had to change directions to get on the board. Preferred footing to change directions for the boards was not always possible because of the water and mud.

Remember, this man going against dad had already driven the two paths countless times as he hauled sand to his future church building. Anyway, dad took one path, and the young man took the other path. There was not any discussion on which path was more difficult. Both were very difficult.

Everyone was having a wonderful time, there was much laughing and screaming even before the wheelbarrow race. The pastor was chosen to start the race. The two racers stood side by side as they were ready to start the race.

Dad was once again dad. He acted like the wheelbarrow was a motorcycle. "RRRuuMM! RRRuuMM!" came the growl from dad. Everyone laughed. Dad tilted his head like he was so serious. Everyone laughed again. "Dad twisted his wrist to make the sounds again, "RRRuuMM! RRRuuMM!" The church's pastor had to ask, "Are you ready, Pastor!" (Dad was his pastor.)

The preacher called "Go!" and they both took off. It was hard to tell if one had an advantage physically over the other. They were both fast and very talented. Dad chose to get his edge by pushing things to an extreme without making a mistake. Dad chose to start and break as hard as he could to maximize speed. He took the turns carefully so as not to miss a step. Dad was slow in turning so that he would not slip and that the wheelbarrow would not dump itself. Dad made sure he lined up the wheel exactly straight onto the boards then he slammed the wheelbarrow into the boards. The wheelbarrow hit the boards so hard that the wheel jumped off the board, and the sand was floating inside the barrow.

Still, the race was close. They were either running, braking, or jumping to the side the whole time. Near the end, dad seemed to be slightly ahead. The young man took a chance by pushing things even harder. One of his feet slipped, causing the young man to lose the race.The young man did not lose. He did what none of the others could do. He proved himself to be an expert with the wheelbarrow.

Everyone congratulated both men. The young man congratulated dad. Dad had already won them over, but he did it again. Dad showed himself to be their servant by working so hard all day.

As far as the rest of us Darnells, we followed dad's lead. We all work as hard as we could. It was an incredible day of bonding and hard work.

4 And we write these things to you, that our joy may be fulfilled.

-1 John 1:4 (WEB)

The Unexpected at the Building of an Auditorium

The Africans were just worn out. They did not have normal endurance for Africans. I could not understand what was going on. They were about to open up a part of their world that I could not comprehend. To this day, I am not sure I understand, but I can tell the story.

The whole church was in the middle of a fast. They did not break the fast for the erection of the building. The fast got in the way of construction. It made them less effective. To me, their choice did not seem logical, why schedule a fast during construction? The church knew that things did not make sense to me. The church's pastor said, "You just don't understand things." I was thinking, "Exactly what things do I not understand?" The church's pastor explained things to me.

> "We have given up our families, our homes, and places to stay, our friends, our jobs, and our future to God, and now we are really sad."

I was thinking, "I would be really sad too if I lost it all." What I did not realize is that I had no idea what was going on. The church's pastor continued, "Since we had lost it all, we now no longer have anything else to give to God." They were fasting because they wanted to give more to God.

As he talked about the price they paid, I realized that I was standing in the presence of spiritual giants. They had lost it all for the cause of Christ. I was amazed by their choices. At the time, I was too young to understand what they were saying.

They had identified one of God's patterns. When people finally give something up to God, He makes them feel so good inside. These Africans had given up and given up and given up to God; God had given them such an amazing high after each gift. They had nothing left to give. They had reached the bottom and they did not want the amazing highs to go away.

They did not fast for regular reasons to fast. They were not trying to be great. They were not trying to impress anyone. They were not notching their spirituality through fasting. They simply had nothing left to give to God, so they gave up food for a while. What else did they have to give up?

I might or might not have told the story about the wheelbarrow. It is pretty cool but relatively insignificant. It did remind the church that dad was for real. They had given up everything to be like dad. What dad did was huge for them. The important part is to get a glimpse inside the relatively new Christians who were already spiritual giants. What these believers lost was huge, but it was only a taste of what they gained. God took their gifts and made them very rich.

29 Take my yoke upon you and learn from me, for I am gentle and humble in heart; and you will find rest for your souls. 30 For my yoke is easy, and my burden is light."

-Matthew 11:29, 30 (WEB)

Dad's Last Stories of His Youth

Patty, Julia, and I wanted dad to tell more stories about his youth before he died. This was long after dad's missionary journeys. Dad was legally blind and could no longer take care of himself without help. Dad rarely told any stories about the past. Even rarer were stories from his childhood. We were longing for a little more of dad before he could no longer talk. This may have been the last time I saw dad when he was still able to tell stories. He is a legend in our eyes. We wanted to know more. We pushed dad, "Tell us more! Tell us more!"

Dad became agitated. We did not realize that we would upset him. Dad simply said, "I don't want to talk about such a terrible time." I did not understand the depths of his sorrows. He was bullied a lot and disgraced by his parent's divorce. I did not understand his sorrows.

Somewhere in his life, he came to a point of giving it all to Christ. Instead of becoming bitter, dad chose to give it all to Christ. I know that I have mentioned this before, but I must revisit stories to tie them back together. Dad had no middle ground. It was either leaning on God, or it was extreme bitterness.

Dad concluded when he was young that he had to surrender to God. If dad did not put things immediately into God's hands, it just did not happen. To stay in the thrill of God, surrender had to come at the first opportunity – not after being worn out by conviction.

> *10 She was in bitterness of soul, and prayed to Yahweh, weeping bitterly.*
>
> *-1 Samuel 1:10 (WEB)*

Why 24 Hours

Why dad ever decided that he needed to challenge the Africans to 24 hours remained a mystery until after he died. The Africans had to choose between 24 hours, later, or never. Before long, God was going to test them and see who they loved more. Were they going to choose God or their families? Waiting to choose is a choice. Waiting is saying, "Right now God, I love my family more than I love you. I choose to avoid the trials and to not trust in You. I am not interested in trusting you for the future." Dad lived – Now is the time to risk it all for Christ!

5 Trust in Yahweh with all your heart and don't lean on your own understanding. 6 In all your ways acknowledge him, and he will make your paths straight.

-Proverbs 3:5, 6   (WEB)

Obligation Free

Those who are being carried along by God's power do not feel a sense of obligation. Those who walk by faith are carried by miracles. The sloshing of God's love and the thrill of the calling are overwhelming motivators. When God's love is so strong within that nothing else matters, motivation is not an issue. Those who have chosen to stay within the splash zone do not need motivation. Those who keep the door of heaven open experience a calling and a love that is bigger than life. A much better word than obligation is compelled. When someone is compelled, more than enough strength invigorates even to the point of fighting it seems pointless.

8 He raises up the poor out of the dust. He lifts up the needy from the dunghill to make them sit with princes and inherit the throne of glory. For the pillars of the earth are Yahweh's. He has set the world on them.

-1 Samuel 2:8   (WEB)

7 Yahweh makes poor and makes rich. He brings low, he also lifts up.

-1 Samuel 2:7   (WEB)

JIM H DARNELL JR

Chapter 12

Solutions to the Ten Problems

Revisiting the Problems in Sharing Salvation

One of the opening themes is that uncommon people find uncommon answers to common problems. Dad found uncommon answers to common problems, and we have a glimpse of how his mind worked. The following is a recap of dad's perspective on the 10 problems mentioned earlier.

Problem #1 - Time/Eternity. Heaven or hell does not begin until life ends, but the vision of God's plan starts immediately. God's grand plan starts with this moment. To live in the wonders of God's love, surrender is now.

4 But God, being rich in mercy, for his great love with which he loved us, 5 even when we were dead through our trespasses, made us alive together with Christ—by grace, you have been saved

-Ephesians 2:4, 5 (WEB)

Problem #2 – Dreams. People often want to do a little living before they accept Christ. They have dreams of greatness in their lives that they do not want to lose. Our dreams do not compare to the wonders of accepting Christ's plans for our lives when we become God's hands and feet. Being God's ambassador is far more satisfying than any human plan.

*9 He will keep the feet of his holy ones, but
the wicked will be put to silence in darkness; for
no man will prevail by strength.*

-1 Samuel 2:9 (WEB)

Problem#3 – Lordship. Lordship is central to Christianity. Lordship is turning over ownership of life to God. It is declaring, "God you lead my life, and I will follow." Staying in the splash zone of God's amazing touch only happens when God is Lord of life. Willing to lose it all to follow God is Lordship.

*7 Don't be wise in your own eyes. Fear Yahweh
and depart from evil. 8 It will be health to your
body, and nourishment to your bones.*

-Proverbs 3:7, 8 (WEB)

Problem #4 – Fear. Fear of hell is a great motivation for salvation and is an effective part of leading people to salvation. Another fear is not being in God's amazing splash zone. When life does not, please God, then guilt, shame, disgrace, bitterness, frustration, and hopelessness replace the splash zone. The fear of not walking in the calling and missing the thrill of God reaching through us as far as God can reach is a terrifying prospect. Fear is twofold: fear of hell, and the fear of replacing God's ordained touch is with conviction.

*23 For the wages of sin is death, but the free gift of
God is eternal life in Christ Jesus our Lord.*

-Romans 6:23 (WEB)

Problem #5 – Guilt trip. The Spirit of God convicts making people feel guilty. Guilt is a wonderful thing. Saying a prayer of salvation only to eliminate guilt is not salvation. The purpose of guilt is to get people to turn from sin to turn to God. When guided by the Spirit, telling people they are hell-bound is the most needed truth at the moment. The focus should be a new life, not just eliminating punishment.

8 When he has come, he will convict the world about
sin, about righteousness, and about judgment;

- John 16:8 (WEB)

Problem #6 – Obligation. An obligation is more of a future-oriented guilt trip, "We should share the Gospel regularly." Obligation relies on personal strength and wisdom rather than on God's power and direction. In the middle of God's will are love, thrill, and a purpose. Those who live in the wonder of God perfectly carry out God's plan. His way is far more satisfying than living under the burden of obligation. The calling drives people to either be obligated or compelled.

36 If therefore the Son makes you
free, you will be free indeed.

- John 8:36 (WEB)

Problem #7 – Fitting-in. Trying to fit in turns the adoration to self rather than God. Fitting in is an incredible craving that all have. The only real fitting in is with God. God gives His all to us, and we give our all to Him. This dual surrender between God and man is the sweetest sense of security and belonging. The most satisfying fitting-in feeling between humans happens when those who have given all to God walk side by side. What an amazing feeling when humans become one in God through loyalty and love.

17 that Christ may dwell in your hearts through faith, to
the end that you, being rooted and grounded in love, 18 may

be strengthened to comprehend with all the saints what is the width and length and height and depth, 19 and to know Christ's love which surpasses knowledge, that you may be filled with all the fullness of God.

-Ephesians 3:17-19 (WEB)

Problem #8 – Notching the gun. Notching the gun is unavoidable, but whose gun is the choice? Notching God's gun is the only real way to live. Those who surrender their will to God give Him all the glory, honor, and praise. God just does not keep the Glory for Himself. God honors those who honor Him. To stay in the splash zone of God's amazing touch, all glory must go to God. The Worthy One multiplies the glory given Him and returns it to honor the unworthy.

8 And being found in human form, He humbled himself, becoming obedient to the point of death, yes, the death of the cross. 9 Therefore God also highly exalted Him, and gave to Him the name which is above every name,

-Philippians 2:8, 9 (WEB)

Problem #9 – Discipleship. Discipleship before salvation is called planting the seed. The focus of salvation was found most effective when seen as an opportunity to enter God's love and calling. Discipleship is guiding the newer Christians down the right path. Discipleship is passing along the vision of God's love and bigger-than-life plan. Discipleship should not be saved until after salvation. Discipleship can be a powerful source of salvation.

1 I, therefore, the prisoner in the Lord, beg you to walk worthily of the calling with which you were called, 2 with all lowliness and humility, with patience, bearing with one another in love,

-Ephesians 4:1, 2 (WEB)

Problem #10 – Semantics. Semantics refers to the meaning of words/phrases. Semantics become a three-fold problem when the words are elevated beyond their meanings, when the treasured words hinder the salvation process, and when the treasured words have a false meaning. People use these Christian catchphrases to hide their sins. Sharing the Gospel is most effective when the hearers hear words that are most meaningful to them. Learning the meaning of words is a good thing, but Salvation should not be linked to saying the right words.

5 Have this in your mind, which was also in Christ Jesus,

-Philippians 2:5 (WEB)

The Solution to the 10

Ten things to avoid are a lot to remember. The good news is the answer is simple. Going in the right direction avoids problems. Surrender always leads people to heaven's throne room. The greater the surrender the more real God becomes. Heaven's greatest gifts come with total surrender. The smaller the reward, the less appealing it is. God offers the most amazing love immediately, and a calling that makes living bigger than life for those who are willing to risk it all and take His journey.

JIM H DARNELL JR

Chapter 13

Persecution

Some are Bent on Sinning

Some have made their choice. Some are going to honor their sins and dishonor God regardless. No matter how high God raises the level of guilt, the choice is the same. No amount of guilt or condemnation from God will deter some from going their way.

The Rebellious Are Out of Control

The rebellious try to find meaning and love in life without God's touch. People always chase after love and meaning in life. God placed within each person a need for love and meaning far greater than human strength to overcome or fulfill. In a desperate search for love and meaning, people make all sorts of bad choices. When God is rejected then the need for love and meaning rages out of control.

Persecution Attacks the Faithful

Bitterness, jealousy, and rage slosh through the rebellious. Nothing they do can diminish the ugliness inside, but they can't quit trying to escape the venom to find peace. As a result, they try to crush, attack, or destroy anything that stands in their way. Those who are bent on going their way, are far from harmless for their cravings are driving them to all sorts of unimaginable ugliness.

Believers who follow God's plans will face severe trials. Persecution comes from outside and inside the church. From the outside is more expected like parents reject their children who come to the Lord. However, when persecution comes from within the church, it can hurt. When someone happens to be in the way of the self-righteous, ugliness jumps out. These people who acted holy suddenly are spewing bitterness, jealousy, and rage. Oftentimes, the believer walking in truth is perceived to be the problem. The gentle faithful is often the target of great persecution just because others have dreams that God will not bless.

31 Let all bitterness, wrath, anger, outcry, and slander be put away from you, with all malice.

-Ephesians 4:31 (WEB)

The Most Faithful are often Persecuted Most

One of the saddest things about the time in Africa is the fate of dad's heroes. Dad chose the men whom God used most to be his heroes. One by one, these families faced such intense persecution. They withstood great trials from without and within the church. By the time we took our last step on African soil 11 years later, all these missionary families had been driven elsewhere.

13 Therefore I ask that you may not lose heart at my troubles for you, which are your glory.

-Ephesians 3:13 (WEB)

Why are those Who Live by Faith Persecuted?

Some people become the symbol of the movement. When some reach the icon status, then they appear as big as the movement. Imagine a believer who lives in the shadows of an icon.

The believer wants to lead, but the voice of the church listens to the icon. Some faultily conclude that if the icons can be controlled, then the whole movement will follow. One poor leadership choice is to indirectly control the movement by controlling the leader. The argument is flawed because icons have tapped into something bigger than they are. The shackling of the icon is not a very good way of leading; but for many, it appears the only way they can lead.

A second inappropriate way of finding a leadership spot is to remove the icon. Sometimes an icon is heavily criticized and all leadership is labeled as wrong. An iconic leader often is faced with the choices of being controlled, cast out, or eventually both.

Anytime someone reaches a legend status, many others around see things the same way. The term legend or icon represents many, not just a handful going in the same direction. It is the one who is most iconic that usually receives much criticism while the others going in the same direction can be easily ignored.

As a child, dad was the least in his community because he was raised by a single mom and was the poorest of the poor. Even those who possess nothing can easily be targets of persecution. One thing is for certain, no matter the status, believers will face incredible persecution which could come from without or within the church.

17 This I say, therefore, and testify in the Lord, that you no longer walk as the rest of the Gentiles also walk, in the futility of their mind,

—Ephesians 4:17 (WEB)

People Headed in Different Directions don't Mix

Those who love to surrender their wills to God, and those

who are out to prove their greatness are going in opposing directions. They have little in common except that they both officially worship the same God. Their theologies clash. They do things for different reasons. God guides one, and the other is wandering recklessly while casting out destruction.

Those within the church who must prove their greatness crave the silent God to bless their work. What they settle for, is that they can gather around themselves and others who need to prove their greatness.

Sometimes their numbers can grow very large, but the blessings of God are far more shallow than desired. They try to prove God's stamp of approval, but His miraculous love and passionate calling don't drive their work.

Some try to take God's work hostage to lead the Almighty's blessing. They try to bump the leader and take over God's blessing. They transfer the shame and discredit that they deserve upon the leader. All sorts of terrible lies, distortions of the truth, and blame are heaped upon the iconic figure.

Are these people saved? Perhaps some are. Some may have genuinely trusted God at some point but have fallen away. Others have said prayers of Salvation with wrong motives like having guilt erased or a chance to start over. It is God's place to decide if the prayer of Salvation was genuine.

The sad truth remains; some within the church seek their glory. Those who do not bear good fruit are in great danger. There is no such thing in the church as a tree that bears good and bad fruit. While God chooses whom He saves, people's fruit is recognizable.

10 Even now the ax lies at the root of the trees. Therefore, every tree that doesn't produce good fruit is cut down and cast into the fire.

—Matthew 3:10 (WEB)

The Perfect Picture of Jesus at the Cross

God is the most iconic figure of them all. Each has played the selfish game which stirs up great wrath within the Father. Everyone has tried to humiliate and shame our Creator and Giver of Life. Each wants to be the god. Everyone has chosen to "Do it my way." My way declares to God, "Not Your way" which is spitting upon God's plan. My way emphatically states, "You are not good enough to be my God. The hostilities toward God have churned up within Him the desire to kill us all. All fell under His wrath and deserved eternal damnation in hell. (This is beginning to sound like a toasty hell, fire, and damnation sermon. Isn't it so fitting?)

God felt rejected and that His plans were spat upon. Anyone who claims, "I am not that bad for I tried to be good" mocks the Maker. Humanity was made to be loved and guided by the Supreme. When the Giver of Life and His plan are rejected, humanity violates the reason for existence. Even the best of humanity stirs the Judge to full wrath because they seek to be their god.

Jesus knew God's anger for He was God Himself. Jesus chose the full wrath of the Father upon Himself. God was incredibly furious and loving both at the same time. Jesus chose to be consumed by the Father's wrath even to point of humiliation, mockery, and death. What Jesus went through is the most unfair thing of them all. The Most Perfect One suffered as the most imperfect.

18 For the wrath of God is revealed from heaven against all ungodliness and unrighteousness of men who suppress the truth in unrighteousness,

-Romans 1:18 (WEB)

Victory #1 Over Persecutions

This is old man style; you have been led right up to the point and you may not have even realized it. If that happened, it is payback for those real old men who did it to me.

The road Jesus took was lonely. He was surrounded by crowds and disciples, but He could count on no one. No one understood or appreciated His plans to redeem mankind. Christianity often focuses on the words, "Well done my good and faithful servant!" Those are great summary words but lack any specifics. The end of all suffering by those who have carried their cross is at the cross of Jesus. Some higher words of praise from the Father are, "Now you know what it is like when Jesus suffered unfairly and gave up His life!" God has an incredible longing for His children to understand His great sufferings. The road God suffered is so lonely, and He longs for someone to understand; that is, someone who suffered greatly unjustly. There is a special bond that God will only share with those who have willingly suffered the same kind of trials.

21 From that time, Jesus began to show his disciples that he must go to Jerusalem and suffer many things from the elders, chief priests, and scribes, and be killed, and the third day be raised up. 22 Peter took him aside and began to rebuke him, saying, "Far be it from you, Lord! This will never be done to you." 23 But he turned and said to Peter, "Get behind me, Satan! You are a stumbling block to me, for you are not setting your mind on the things of God, but on the things of men."

-Matthew 16:21-23 (WEB)

What is the Difference between Guilt and a Judgmental Attitude?

It is easy to feel justified when others are judged. When the sins of others are stared at, judging can appear to be so justified. Staring at other people's sins enflames condemnation. Dwelling even just a little bit on the sins of others inflates amazement of failure. A "righteous indignation" feeling explodes within those who are judging.

Guilt and judgmental attitude seem so different; but in a way, they are the same. Guilt is the Spirit's declaration of wrong. A judgmental attitude supposedly declares both the failure of others and self's faithfulness. Exalting self only feeds the judgmental attitude. If someone has to be more, then others have to be less. The two ways to increase the greatness gap are to exalt self and judge others. The drive for greatness envisions the self as big and others as little. The root of righteous indignation is an attitude of being a super self.

God places His pointed finger on pride. No matter if the prideful feel like their actions were good or bad, they burn. If they feel like they did misbehave, then guilt consumed them. If they feel like they did well, then righteous indignation burns toward the sins of others. God touches the self-righteous proud with a flaming burn no matter what they do. Did I just chase a rabbit or does this all fit together? Judging is an attempt at transferring righteousness to self and guilt away to others.

1 "Don't judge, so that you won't be judged. 2 For with whatever judgment you judge, you will be judged; and with whatever measure you measure, it will be measured to you.

-Matthew 7:1, 2 (WEB)

14 "For if you forgive men their trespasses, your heavenly Father will also forgive you.

-Matthew 6:14 (WEB)

Spiritual Warfare

All troubles are spiritual warfare. The devil seeks to destroy believers while God seeks to refine them. The Africans were right. All troubles are a result of the spiritual fight. Unjust wrongs, failure, guilt, and judging are all a part of spiritual warfare. Labeling troubles for what they are is very important.

Victory #2 Over Persecutions

Warfare is all about winning. How can a war be won when the enemy is not identified? The enemy is not persecution, being treated unfairly, or the viciousness of others. Fighting the visual always leads to catastrophic loss. Who wants to fight and lose an arm or a limb or even life itself out of ignorance? The spiritual loss of worry, wrath, bitterness, jealousy, judging, despair, and guilt is far more destructive than the loss of a limb.

Victory 2 comes in winning over worry, wrath, bitterness, jealousy, judging, despair, and guilt. Brokenness will come, but victory rises in humility. With great troubles comes an opportunity for victory. Troubles seem to be something to overcome. Persecutions are things to let God shine through. Troubles seem mundane, but persecutions are visualized as great warfare, but both are great warfare to have triumphed over with the meekness of God

32 And be kind to one another, tender-hearted, forgiving each other, just as God also in Christ forgave you.

-Ephesians 4:32 (WEB)

Victory #3 Over Persecutions

Troubles easily make people feel like failures. Sometimes troubles end in a disaster where the entire world seems to see the victim as a failure. The natural response would be, "If only I could have somehow done a little more, then I would have survived!" The mind races non-stop trying to figure out how to be better so that something like that would not happen again. Such reasoning is limited in value for overcoming life's problems is not the main point. God has His plan that cannot be thwarted. How wonderful it is when God finally speaks, "If you had been better, I would have made the test bigger!" Nothing is more honorable than to lose an outward battle than to win a greater inward battle. Even if everyone sees a great outward loss rather than a great inward victory taking place, winning the spiritual battle is always so worth it. The madness of such thinking is obvious, but great spiritual victories always come at a monumental price. In times of great troubles, the need is not to overcome earthly attacks, but to win the spiritual battle. Those who are willing to lose much can win much. Great humiliations need to be welcomed in peace so that the real battle can be won.

10 Then many will stumble, and will deliver up one another, and will hate one another. 11 Many false prophets will arise and will lead many astray.

-Matthew 24:10, 11 (WEB)

2 Count it all joy, my brothers, when you fall into various temptations,

- James 1:2 (WEB)

Nowhere Else to Go

The picture of that wonderful man comes floating. The man said, "Pastor, I have nowhere else to go." The man who lost it all because He stood for Christ. The man was so broken because he lost his family, his home, his livelihood, and his future. Without a miraculous intervention, this man was dead shortly. This man, after tasting the goodness of the Lord, wanted nothing else. He had nowhere else to go because he chose God.

19 Go and make disciples of all nations, baptizing them in the name of the Father and of the Son and of the Holy Spirit, 20 teaching them to observe all things that I commanded you. Behold, I am with you always, even to the end of the age." Amen.

-Matthew 28:19, 20 (WEB)

9 He has said to me, "My grace is sufficient for you, for my power is made perfect in weakness." Most gladly therefore I will rather glory in my weaknesses, that the power of Christ may rest on me.

-2 Corinthians 12:9 (WEB)

Chapter 14
Revival

What is a Revival?

Revivals are when God's Spirit has come down upon people with great power. God sloshes His love, and His calling is so evident. Incredible fear seizes the faithful; they are afraid that they might step out of the incredible downpour of love and the amazing journey of being God's hands and feet. The faithful chose to risk it all to walk with God.

Revivals are marked by intense surrender to God and amazing blessings from above. Some have characterized revivals as a time of great repentance. Revivals are much more than just an overpowering release of guilt. Revivals are turning from all known sins to relying on God instead. Relying on God is surrendering the will to Him. No matter how revivals are sliced and analyzed, it always comes down to the same thing; a group of believers fully rely on God with a very powerful response from the Almighty. They have chosen to forsake it all.

10 that you may walk worthily of the Lord, to please him in all respects, bearing fruit in every good work and increasing in the knowledge of God,

-Colossians 1:10 (WEB)

When do Revivals Come?

Revivals always come when believers chose to forsake all for

Christ. Anytime a group chooses to live in the slosh zone of God's love and chose to risk all for Him, revivals always come. There is no maybe. Those who take the steps to enter into revival always enter into revival. God does not show favoritism to some groups.

> 18 Don't be drunken with wine, in which is dissipation, but be filled with the Spirit, 19 speaking to one another in psalms, hymns, and spiritual songs; singing and making melody in your heart to the Lord; 20 giving thanks always concerning all things in the name of our Lord, Jesus Christ, to God, even the Father;
>
> -Ephesians 5:18-20 (WEB)

Revivals are Cool

It is so amazing to be in a room full of people who would so eagerly give their lives one for another as God calls. The bond is so intense that the feeling is beyond description. God's love just seems to be bouncing from one person to another in such an incredible frenzy yet at the same time still fully being in everyone. Such love was the norm, not the exception. When all things are surrendered, God speaks love. Love is so intense that it is more than a feeling. Actively loving others is a must. Love cannot be contained, held down, suppressed, or even wait. God's love knows no borders and it must go.

Growing up, I thought such love and devotion were the norms. Sadly, I have also lived among many who refuse the cool winds of revival by killing the love flow. Still, I have seen and tasted the revivals. God had chosen in His great love for me to experience far more revivals than I could count. I feel so unworthy of God's wonderful revival gifts to me. The revivals compel– Be like a revival!

Gotta Go!

"I gotta go" means I gotta go now. You know the feeling; the legs start rocking back and forth as the urge increases. Eventually, the legs fold up and twist like pretzels. The legs squeeze so tightly trying to hold it all in. After a while, the urge to start jumping up and down overtakes. Now jumping up and down is a natural response to distract from the all-consuming urge; but logically, the jumping would seem to serve as a tool to shake it out. The jumping is not normal jumping for the legs somehow still stay twisted like a pretzel. The body and mind are fully concentrated on not springing a leak.

I gotta go seems like a terrible analogy for a revival, and perhaps it is. At the same time, I gotta go perfectly describe a revival in intensity. Those in revival gotta go because love urgently calls to action. Those in revival see God's mission all around. God's love drives the calling.

There was a story that I knew God wanted me to include. I wrote the book and did not find a place for it. As a child, I did not understand the story. I knew it meant something, but The Revealer of Secrets withheld the treasure. In the editing, God said, "This is the place!"

Many Africans did a spiritual pilgrimage as I saw it then. They would set aside the day in advance. All their thoughts leading up to and beyond were completely focused on the day. I remember Africans saying, "On Saturday, I will stand by the street corner and preach from sunup to sundown." They had planned to preach for 12 hours straight. They would pick the busiest street corner and stand in the way. During daylight, they would preach Salvation to anyone who would listen. On the way to work, people would hear; and on the way home, people would hear again.

The time would vary greatly: in six Saturdays, in three months, or next year. The date was often set way in advance for today such an adventure was not possible. To make the impossible possible, much time was needed to wrestle with God and learn.

I thought the plan was terrible. I had passed by street preachers. Most ignored them. A few walked way around them. Some were hostile. I was too young to endure the lack of response. Had I stayed, I would likely have caught the bug. I would have been all consumed as well. How could I continue to preach my heart out so with the enthusiasm that every sentence was the most important sentence and do that for 12 hours? I would never know in what few random seconds when a life would be changed forever. How could I allow the intensity to tone down because I did not feel like it just then? What life would be less important? Every moment for 12 hours needed to be the most important ever. No amount of apathy from the crowds could discredit one moment.

I remember watching a street corner preacher once. He had the Bible in his left hand. His right arm would flail upward only to zoom down and thump the Bible. He was almost screaming for he wished that the whole crowd could hear. He spoke fairly fast so that every ear would be captured by the plan of Salvation. His voice popped with intense conviction. The crowd was so thick that they had to brush against him as they passed by.

I remember passing him by, and only in a few moments, he was gone. I looked at him, and he at me. I did not know what to say. I have always wished that somehow, I had cheered him on. I tried to find the words, but nothing came. A terrible silence imprisoned me that day. It was too strange for me to grasp, yet his words have once again been released. His message was similar to this, "God is calling you to make peace with Him. He has a wonderful plan. Repent so that God's love will come in." He finished preaching after 12 hours, but 30 years later, his words are still echoing. My calling was not to say anything that day, but this day. If I had felt like I had done my part, I may not have remembered the moment. Today, I cheer him on. He won because He let the Molder prepare him for such a glorious day. Today, the Whisper said, "You know understand. He had to go because of my love."

I am picturing myself as an 11-year-old boy who could not have possibly preached for more than 20 seconds and that included stutters. I lacked the strength to boldly stand on the corner all day. I could not say anything with that kind of conviction. Being a street corner preacher was impossible. I imagine God could have questioned me, "Do you love Me enough to stand on the street corner and preacher your heart all day?" I would want to say, "Yes!" But I couldn't. Lip service is not an acceptable answer. There are only 2 answers, "Yes, I love you enough to stand on the street corner and preach all day!" or "I don't love you enough to stand on the street corner and preach all day!" Saying no would be walking away from a real commitment to God. Saying yes would consume me. From the first yes, the training would begin. God would have to teach me how to be brave. He would need to loosen my tongue by teaching me how to speak. He would have to break my shell so that I could proclaim. Making good eye contact would fall in there somewhere too. Such a love journey would become a pilgrimage.

Common people choose to stand on the street corners and preach. That is what God's love does – it drives people to make the most intimidating pilgrimage. Revival love is all-consuming. Can you imagine a room full of people where everyone has an impossible pilgrimage set before them? Together they fan the flames of love and the calling where the revival is in full bloom.

The Lost Cause

Trying to start a revival is a lost cause. Revivals cannot be started; they can only be caught. I have heard countless preachers, speakers, and singers say, "May a revival begin, and let it start with us!" Revivals are magnificent. Being a part of the spark is so amazing, but lip service is a waste. Such revival talk often means something like this, "Let's get so excited about God so that everyone will catch the frenzied excitement, and let it be because of me."

When was the last time you ever saw someone so excited about nothing? Imagine a person who is so excited about nothing at all that there is shouting, singing, and dancing? You ask, "What are you so excited about?" The genuine answer returns, "I have no idea why I am so pumped but isn't it exciting?"

Romantic love is a different story, "She looked at me! She knows I am alive!" That young man is all giggly, and says, "What?" "What?" all day because he cannot concentrate on anything else. For a real cause, people get excited.

Revivals have always been started by people who found the Real Cause. The Real Cause is found through extreme sacrifice of personal will. God is not looking for people who seek to be the ones. God is the only One. God is calling people to become nothing and fully empty for God's love and calling need space to fill. Revival is a biproduct of full surrender and awakenings can never be achieved directly.

How many does it take for there to be a Revival?

When God moves with power, there is a revival. One of dad's favorite phrases is, "God and I are a majority!" Dad lived in the thrill of a revival even when walking alone.

There is this truth, "A believer cannot disciple others to be more spiritual than they are." People can only show the path that they are on, yet God speaks as well. Some disciples outperform the masters. God in His grace takes the master's words and adds far more. God can train a disciple to walk more faithfully than the spiritual mentor.

Mom and Dad journeyed to lands that had no understanding of God. Mom and dad spread the Gospel, and revivals broke out. So here is the question for you, when did the revivals start? Did the revivals start when God seized the people with power? Or were the revivals just spreading from a common source? Maybe God used a poor source and added much more? These questions are asked not to honor mom and dad, but to better understand God's spark. If lots of revivals started from the same source, then the source was already in revival.

All definitions that I have ever studied or heard have said that revivals include lots of people. There is a hypothetical question that has an exciting solution. Can a revival be limited to two people like mom and dad? Let's press this idea a little further. Sometimes, dad's journeys took him away from the family, and revivals sprung up. The source at these times was just 1 person. Answer this question, can a revival be just one person?

It is time to start nit-picking. Sometimes mom and dad or just my dad found themselves in places where the harvest did not come. If my parent(s) reached out with the same passion and message, and nothing happened is there a revival? These are difficult questions that need to be posed.

The answer to these questions can be thrilling. God can choose to bless just one like He blesses the multitudes in a revival. Sure, the road is lonelier, but the grandeur of God can be just as real. Revivals refer to the condition of the heart that God abundantly blesses. Any one person can live in a state of revival. God can bless one person just in the same manner that God blesses lots of people in a traditional revival. Revivals are far more than just events; they are a way of life.

20 Now to him who is able to do exceedingly abundantly above all that we ask or think, according to the power that works in us,

-Ephesians 3:20 (WEB)

Revival of One

A revival is not measured by large numbers. Revivals are measured by an eagerness to surrender. Those who choose to walk with God will be persecuted. The troubles carve out the revival gem for the darkest of times have a way of erupting into revivals. Before the revival is seen without, it must be within. Troubles can whip and life can seem pointless, but that is where all good stories go.

Chapter 15

The Movie

The New Believer Goes Home

There is another African man that became my hero and a hero to much of Africa as well. A man heard an American missionary talk about the wonders of God. This man was so convicted about what he had heard about Jesus, that he went home to tell his family.

Upon arrival, this man told his family about Jesus. The response greatly saddened this young man. His whole family rejected Jesus. However, this father did not kick his son out of the family. This man had other plans – very sinister plans.

From the father's perspective, this was a spiritual war. It was a war between spirits the family had worshiped through endless generations against one man and his God. The father had schemed a great plan to win his son back to the spirits that their fathers worshiped.

The father visited the witch doctor, marabou, and explained that he needed help in terrifying his son. The witch doctor knew just what to do. He gave the father two powders. One powder was a poison, and the other powder was a cure. Now the poison was not deadly, but the burning was bad enough that people who take it would beg for mercy and wish that they were dead. The poison would wear off in a few days and leave no permanent damage. Almost immediately after the cure was taken, things would go back to normal.

The plan was for the father to spread the poison in the dirt by the gate. As the son would walk through the gate, he would be infected by the poison. The son would then beg the father for help. The father would explain that the son is sick because he rejected the spirits. The father would only give the healing potion to the son upon rejecting Jesus. That was the plan.

The father sprinkled the poison in the sand by the gate. The whole family swallowed some of the antidote. The family had gathered to watch the son as he walked through the gate. Soon enough, the son passed through the gate and saw the family staring at him. They kept staring at him waiting for the poison to take effect. The son was staring at the family all confused not knowing about the poison. The family was getting frustrated that nothing was happening to the son.

Suddenly, the whole family became sick at the same time. They were screaming and wailing like their insides were coming out. The son was standing there not understanding why the whole family had been staring at him and are now so sick.

Between uncontrollable screams in pain, the father told the son the story. The witch doctor had given us two powders. One was a poison, and the other was a cure. The poison is not deadly but would fade in a few days. Until then, whoever took it would wish to be dead. Mistakenly, they had poured the cure in the sand, and they had swallowed the poison.

The father and the family were crying out to the son for help. Of course, the son could not help because he did not have the cure, but the son consoled them. The son reminded them that they would feel better in a few days.

The father confessed, "This was spiritual warfare. Your God proved greater than my gods. I reject the spirits that our fathers had worshiped. From now on, I worship your God." The father chose Jesus.

The son's sorrows turned to joy, for his father met the Redeemer. God did what He always did. He snatched away the guilt and the fear of eternal doom. This man forced a showdown and lost, yet they still somehow won the greatest prize of all. He lost God's judgment and gained wonders. The Loving Father wrapped His loving arms around the family and reaffirmed, "You are mine forever!"

The son made a choice. After just meeting Jesus, he returned home. The son knew that the family would likely reject him. He knew that he would likely face intense persecution – persecution even unto death. The son chose to lose it all because he had no other choice. If he tried to keep the treasure of God all to himself, he would lose it. He had no choice but to tell and risk losing it all.

Here is what God did for the son. God gave his family back to him. In a matter of hours, the son went from being lost to finding it all, to losing everything that had mattered to him to gaining his father back. This son became a spiritual legend to his village and Africa.

Africa? Oh, yes! You see, his story so inspired my dad and Ed Pinkston. They decided that a movie had to be made. I remember the days of them trying to figure out who the actors would be.

15 Having stripped the principalities and the powers, he made a show of them openly, triumphing over them in it.

-Colossians 2:15 (WEB)

The Influence of the Movie, Le Combat

Once the movie was made, Dad and Ed would show it when they were starting a new church. This is the only movie that they would show. You can now see the story circling back around. Many more would accept Christ because of the movie than without it. The movie was incredibly successful.

The movie's fame spread far and wide. Missionaries of many denominations who lived hundreds of miles away heard about the movie's impact. The missionaries made a point, "Why should you be the only missionaries that can show the movie? Why should we be deprived of God's blessing?" As a result, the movie was released so that missionaries of other denominations could show it.

One new Christian endured persecution, and the Rewarder was abundantly blessed. His father trusted Jesus. This man was not just given his father, but families all over Africa. Whole villages and families put their trust in Jesus as a result. People were craving God rather than the wicked spirits.

11 Behold, we call them blessed who endured. You have heard of the perseverance of Job, and have seen the Lord in the outcome, and how the Lord is full of compassion and mercy.

- James 5:11 (WEB)

Chapter 16

The Church Is Born

68 "Blessed be the Lord, the God of Israel,
for he has visited and redeemed his people;

-Luke 1:68 (WEB)

New Church – New Challenges

There were some things that dad could not do. Everyone was wanting his attention. He was not free to organize the new believers into a church. Mom did that.

The second Sunday included Sunday School. Somebody had to organize Sunday School. Most had never led, and none knew anything about the Bible. They could not teach the Bible because they did not know anything.

The challenges always varied depending on the numbers and ages. My mom would be left with teaching several classes all at the same time. On the first Sunday of a new church plant, mom would choose volunteers to assist in the teaching. The volunteers were always overwhelmed and terrified. Many of the recruits were considered too young to count as leaders within their communities. I remember mom picking teenagers to teach the children. At some church sites, mom could only find one teenager to teach a large number of children. Mom had to use what God provided.

Before Sunday School on the second Sunday, mom would explain what they would do first. Mom would go to one group and get them organized and started. The leader would keep things flowing. Mom would move on to the next group, start the lesson and get them going. She would leave them and perhaps work with still a third group. Mom would get every group going and let the leader keep things flowing.

Mom would then return to the first group. By then the activity was running dry. Mom would start the second activity and get it going. She would turn it over to the leader and leave. The second group needed more direction. Mom would return to them and breathe more life into the spiritual learning. By the time mom could free herself of the second group, the last group had just become in need of her help. Mom rotated through again and again until Sunday School had terminated.

By the end of the first Sunday School, something became apparent. Some leaders were not ready to lead, and others showed skills to lead. Some came for the first time on the second Sunday, and mom recruited them to teach. Those who were not ready to lead were glad to let someone have their spot.

Somehow, I never saw a group that had to stall out for a lack of direction. I saw many groups come to the point where they did not know what to do next without direction as mom arrived. I watched in amazement as mom could teach several groups all at the same time and train the leaders. Mom always had enough leaders, but the leaders were very green.

49 For he who is mighty has done great things for me. Holy is his name. 50 His mercy is for generations and generations on those who fear him.

-Luke 1:49, 50 (WEB)

Continuing Sunday School

You would think that the third Sunday would be easier than

the first two Sundays, but it was usually just as hectic for mom. Every Sunday more and more people came.

It is true that as leaders gained more experience, they needed less help. Still, mom had to check in on them regularly. The leaders, even after a couple of months, were still new Christians without a Christian foundation. Many grew so fast in the Lord that their leadership was outstanding.

Mom was constantly having to recruit new teachers for new classes. Especially as the children's departments grew, mom's time was taken. Children require a lot of attention. Mom would have to teach leaders how to manage children. The teaching skills that mom taught the leader the Sunday before, did not apply very well when the numbers doubled.

The new church needed to be fully organized before it could be on its own. Dad was doing his part and mom hers. They were both so busy on Sunday. Together, they organized new believers into a church.

76 And you, child, will be called a prophet of the Most High; for you will go before the face of the Lord to prepare his ways,

-Luke 1:76 (WEB)

Six Months or Less

After a while, the new church was ready to stand on its own. Mom and Dad had gradually turned over all the duties. Dad would draw a man from another church and appoint him as pastor. The church was ready, to stand on its own. Most church plants took three months. The numbers kept growing rapidly so that people with so little understanding, yet such great faith became spiritual giants almost overnight. The fans of revival were blowing.

77 to give knowledge of salvation to his people by the remission of their sins 78 because of the tender mercy of our God, by which the dawn from on high will visit us,

-Luke 1:77, 78 (WEB)

Chapter 17

Pastors and Missionaries

Seminary?

The traditional model of pastor development was to send pastor candidates off to seminary. There was a French seminary in West Africa. Dad realized that the cost of seminary was outrageous. Outrageous compared to his budget. Dad would have to pay all the expenses for the men sent. There was no guarantee that the trained men would ever come back to work with dad. Seminary-trained men were in high demand. Dad did not know what to do.

I remember dad and Ed Pinkston would spend many hours talking about the problem. Both men were having the time of their lives, spending themselves in God's ministry. Dad realized that it would be far better for them to train their pastors. They had a problem. Ivory Coast spoke French, and seminary materials were scarce. Dad heard that there were some in the Caribbean. Dad sent a letter requesting permission to use their materials. The authors responded by giving dad permission to make copies in use for pastoral training.

Just like that Ed and dad became professors. No one ever called them professors, but that is what they became. Dad started teaching classes at night. Dad would travel to one of the churches and teach.

9 For this cause, we also, since the day we heard this, don't cease praying and making requests for

you, that you may be filled with the knowledge of his
will in all spiritual wisdom and understanding,

-*Colossians 1:9   (WEB)*

Bold

The Africans were very bold. African culture had a way of eliminating shyness. They were just bold people. Anything the Africans were asked to do; they would do. Not understanding the task was not a problem. The Africans just chose to step up when called upon.

My dad had a joke. He would say, "If you asked an African to give a speech on aerodynamics, he would get up and give the speech without having any clue what it was." When asked, the Africans would step up. Their courage was incredible. Even if their culture demanded boldness, stepping up was a real act of courage.

At the start of ministries, many people had to be placed in positions that they did not have the skills to accomplish. Usually, they consented and took the job. Amazingly, the churches just flourished and multiplied.

3 In the day that I called, you answered me.
You encouraged me with strength in my soul.

-*Psalms 138:3   (WEB)*

Pastoral Training Class Attendance

When dad offered a pastoral class, everyone thought that the class was for them. People just showed up. Dad never had the problem of having a small class.

Dad picked locations where some of the other churches could attend as well if they could find transportation or be willing to walk for a very long time. Dad would also give some of the men bus money. Dad would teach several pastoral training classes at the same time.

Churches kept requesting of dad, "Teach us a class." I remember one day, the church at Yopougongare (we will talk more about this church later) put in a request, "We have so many people wanting to take a class, will you come and teach 2 classes?" Their building was not big enough to hold all the candidates at one time.

5 that the wise man may hear, and increase in learning; that the man of understanding may attain to sound counsel:

-Proverbs 1:5 (WEB)

Picking a Pastor

In the classes, dad would get to know the students fairly well. Now and then, someone would shine and show much promise as a pastor. Dad had already figured out, who was the readiest to become a pastor.

When a new church was ready to stand on its own, Dad would pick a pastoral candidate and ask the young man if he was ready to pastor. This was always a big shock to the young man. Even the students in the pastoral training classes were very young in the Lord. There were no long-established churches from which to pull candidates. The candidates came from the churches they had planted not long before.

I remember one time when dad asked a man if he was willing to pastor a church. This man started hyperventilating. I thought he might pass out. I felt sorry for the man. The man asked, "What will I get paid?" Dad answered, "Nothing!" The man asked, "How will I make a living?" Dad answered, "I don't know?" Making a living in Africa at the time was extremely difficult. There were no jobs to go around. In Abidjan, the main city in Ivory Coast, the unemployment rate was 50%. Everywhere people were looking for work.

Dad tried to cheer this man up, "Tell you what, I will double your pay." The man laughed knowing the sum was still zero. Dad wisely said, "Take a few days, to decide. Pray about what God wants you to do. Just don't take too long, I will need to ask someone else before long."

A few days later, the man promised dad that he would take the pastorate. Dad kept in contact with this man for a while until he settled in both as a pastor and financially.

> 4 The Lord Yahweh has given me the tongue of those who are taught, that I may know how to sustain with words him who is weary. He awakens morning by morning; he awakens my ear to hear as those who are taught.
>
> -Isaiah 50:4 (WEB)

Erecting a Building

Dad always provided new churches with a place to meet. He would always wait until they were established to find a permanent location. Dad sometimes helped them build a place. Dad became a good general contractor. Like everything else dad did, he trained people to take as much of the load as possible.

> 4 One thing I have asked of Yahweh, that I will seek after: that I may dwell in Yahweh's house all the days of my life, to see Yahweh's beauty, and to inquire in his temple.
>
> -Psalms 27:4 (WEB)

Other Missionaries

Abidjan, where we lived became a hub for ministry support

personnel. They started publishing and creating media outreaches for all of West Africa. These missionaries came with specialized skills and were not pastors. Dad tried to find ways to include them in his ministry. Dad connected these missionaries to churches. The plan was for these missionaries to mentor these relatively new churches. Dad also hoped to free himself up a little more so that he could focus more on new work. Dad hoped that this would greatly benefit all involved.

13 Take firm hold of instruction. Don't let her go. Keep her, for she is your life.

-Proverbs 4:13 (WEB)

New Church Planters

When new church planters came on the field, dad would take them with him and let them experience how dad started churches. Before long, the new missionaries were quite successful in planting churches of their own.

10 Receive my instruction rather than silver, knowledge rather than choice gold.

-Proverbs 8:10 (WEB)

Missionary Houses

At one point, there was a large influx of missionaries. Not enough houses were available for all the missionaries. The housing market was outrageous, and dad was allocated a modest sum to buy houses. Dad was stuck, there was not enough money to buy, and the missionaries would soon be there.

Dad bought three side-by-side lots an hour away. Dad built 2 houses. We ended up living in one of those houses. It was a wonderful place to live. I left all my African friends behind for my parents never took me back to my former neighborhood. I never saw them again.

1 Unless Yahweh builds the house, they who build it labor in vain. Unless Yahweh watches over the city, the watchman guards it in vain.

-Psalms 127:1 (WEB)

Chapter 18

The Caretaker

D ad had hired a man to be caretaker for the three lots. Even after the houses were built, there was still a lot of outside work to be done. This man loved to work. He would race around as if he were competing against someone. He never slowed down. He would sing loudly as he worked. In all his enthusiasm, he never could hit a note. This man displayed his love for the Lord in the way he worked.

Two dark events were looming on the horizon, and neither was expected. Together, there was no escape. I probably would not remember the man except he had won over dad's heart and mine. Dad was usually persistent enough to acquire what he wanted, but this was an exception.

Africans were very afraid of evil spirits. They had all seen the power of witchcraft do some terrible things. They also knew that some spirits were fake while others were real. The only way to tell the difference was in the power displayed. The power always had one thing in common; the spirits cursed and destroyed people.

A new spirit scare had erupted. Anyone who touched a man from a certain tribe would be cursed. The general population was mortified to touch anyone from that tribe. Mortified is a good word and not an expression. They thought that they might die if they touched the tribe.

The caretaker was working outside as usual. One day, he stepped on a nail that pierced the bottom of his foot. This is the tropics, everything metal rusts fast because of the humidity. The caretaker went to tell dad. Dad was deeply concerned and took the man to a medical center. The people in the center were very concerned as well. Rust often carries tetanus, which is soon deadly left untreated. The medical personnel looked at the wound, it was all puffy and red. They knew he was most likely infected with tetanus, and it was only a matter of hours before the man died. Still, the medical personnel were more concerned about the scare than about this man's life. You see, no one had to ask the caretaker what tribe he belonged to for that tribe had some distinctive facial features. Dad was a very persuasive man, but he could not convince anyone to give him the shot that would save his life.

Dad always had a backup plan. Dad simply took the man to another clinic. At the second clinic, dad found the same problem. No one was willing to give this man a shot. To be honest, shots can be administered without touching, but the fear of being cursed by the spirits was overwhelming. Dad went all over town trying to find a clinic that would give him a shot, but no one would. As the day progressed, a red streak was running up his leg climbing ever higher. Dad had one last plan.

1 Praise Yahweh, my soul! All that is within me, praise his holy name!

-Psalms 103:1 (WEB)

Were my Teachers Old Men in Disguise?

I remember back in elementary school when my teachers would read books to us. When they got to the best parts, they would stop and say, "We will continue reading tomorrow!" We would all say, "Aww!" We just had to hear a little more to see how things turned out.

All my elementary teachers were women, but I am sure they must have had a little bit of an old man inside. They just had a way of torturing us because they did not do stories the normal way like continuing with the story of a man's life which was slipping away fast.

Military Service

Dad spent time in the Marines. They would line the men up and give them shots. Shots in those days were quite different than shots today. Today, it is possible to get something like a 6 in 1. That is six different medicines in one shot.

When I was young, we got 1 in 1. The shot was for one disease only. They had not mastered shots back then. The shot designers had no clue how much medicine was needed to keep the disease away. Would giving a little more medicine make the shot last longer before a booster was needed or even keep people from dying? The experts had to guess how much medicine was enough. They always erred on the side of caution. They gave new meaning to the phrase, "It is better safe than sorry!" After the shot, everyone was sorry they had gotten one. The shots were often too strong. I wonder if this was the litmus test, "Look! That person is still standing after getting the shot, the dosage must not be strong enough!" I know that this is a little bit of an exaggeration. We all had permanent scars from some shots. Once my arm became so swollen that I could not bend my elbow hardly at all.

Back to dad! Dad was a real man, and he proved it one day in the clinic. The technician had inserted the needle into dad's arm. Sometimes, the medicine burns. On this particular day, the medicine got the best of dad, but just for a moment. Dad flinched! You would think that the needle would tear the muscle during a flinch. This time, the needle was lost. When dad flinched, the needle snapped. How can you have a more glorious military career during peacetime, than to snap a needle while getting a shot? Okay! Okay! The commercial is over!

The Caretaker Continued

Dad had one last plan. He realized that no medical personnel was going to give the shot. Remember, shots made dad very squeamish. Dad asked for the shot so that he could administer it. Dad injected the caretaker with the shot. It would take a few more hours to see if dad was in time.

It was expected, perhaps at first, that the caretaker would get worse until the medicine had a chance. The red streak continued to grow. Even after a little bit, the caretaker continued to become worse. Before long, it became evident that the shot had come too late. The man only lived a few more hours and died.

I tried many times to make some sense of this story. Perhaps now, I have a better understanding. Sometimes, God waits seemingly a lifetime to give answers.

First and foremost, this man is a hero. He is not just anyone's hero – he is God's hero. This man served God with all his might. When he worked, he worked unto the Lord. In his mind, he was not working for my dad; he was working for his Savior. The hard work was a gift to God, and God was more than pleased with the gifts of labor. God filled this man with incredible joy. Even unto his death, he was filled with joy. Dad held the man close as he died. The man did not whine or try to make peace with God. Everything was settled beforehand. Life or death was not important. Loving God and being loved in return was all that mattered. This man is God's hero. He proved he was a hero to the very end. He is a hero for many others as well. He was one of my dad's heroes. He blessed dad every day with the joy of the Lord. When dad told me the story, he was greatly moved. I was away at boarding school at the time of the caretaker's death.

He is my hero and perhaps yours as well. He proved that one could die in the joy of the Lord even when the world curses you. God gave me an incredible gift to have been touched by this man. If this story touches you the way it touched me, then he is your hero as well, and God has given you the incredible gift as well. This man is a real hero.

> 21 "His lord said to him, 'Well done, good and faithful servant. You have been faithful over a few things; I will set you over many things. Enter into the joy of your lord.'
>
> -Matthew 25:21 (WEB)

For Every Man, a Time is Appointed

Long before the caretaker ever was, God had appointed the day and time that this man would die. The world was not spinning out of control. God had ordained the moment He would send His angels to collect the precious soul.

> 27 Inasmuch as it is appointed for men to die once, and after this, judgment, 28 so Christ also, having been offered once to bear the sins of many, will appear a second time, without sin, to those who are eagerly waiting for him for salvation.
>
> -Hebrews 9:27, 28 (WEB)

Cursed?

What about the curse? The Western mind says that all things happen by natural causes. This man died because he did not receive medicine in time. Had he received medicine in time, he would have lived. The African mindset is that anyone who touched someone from this tribe would be cursed. They honestly believed the curse was the truth. So, what is the truth? Well, neither. My dad touched this man, and nothing happened. So, it could be assumed that the curse was a fake, but that is not what happened. The curse was real because it killed a man. Demons seek to kill and destroy for that is their mission. The best way to destroy a great spiritual warrior is to discredit him with a curse. Evil spirits chose a plan to kill that great man of God. Evil spirits are liars. Would they not rather humiliate and discredit a man of God before killing him?

> 12 There is a way which seems right to a man, but in the end, it leads to death.
>
> -Proverbs 14:12 (WEB)

Who Won?

Who won – God or the evil spirits? Is that even a fair question? God won all the way. Two things happen at the same time: evil tempts and God's test. The caretaker rejected being tempted to fear for his life. The caretaker rejected being bitter at the medical personnel. This man of God neither let his soul be poisoned in bitterness nor did he lose focus on the joy of the Lord. The caretaker passed God's test proving incredible love and loyalty to God. God and the caretaker won.

> 12 but he, when he had offered one sacrifice for sins forever, sat down on the right hand of God, 13 from that time waiting until his enemies are made the footstool of his feet.
>
> -Hebrews 10:12, 13 (WEB)

Heaven's Jubilee

God never stops being excited about his plan. Even when troubles seem to be all around, God is excited. The Maker is thrilled because He sees the beauty of His purpose. At any point, if God ever grows tired of the humanity game, he could end it. God has chosen to spend His eternal time with men.

Even in the darkest moments of our life, the thrill is heaven is beyond words.

> *2 looking to Jesus, the author and perfecter of faith, who for the joy that was set before him endured the cross, despising its shame, and has sat down at the right hand of the throne of God.*
>
> *—Hebrews 12:2 (WEB)*

The Power of Evil Spirits

Evil spirits are far stronger and more terrifying than any human can bear alone. These spirits destroy and kill. They are very effective in leading so many to hell. They should never be taken lightly. Part of the Western curse in cause and effect is that they pretend that the spirits are not eager and powerful enough to destroy. Westerners often go through life not taking spiritual matters seriously. When they reach their appointed day, they realize how wrong they were.

> *8 Be sober and self-controlled. Be watchful. Your adversary, the devil, walks around like a roaring lion, seeking whom he may devour.*
>
> *—1 Peter 5:8 (WEB)*

Godly Focus

Some Christians spend their lives trying to fight and rebuke evil spirits. They correctly see much evil around, but they are misguided. Those who stare into evil do not experience the joy of the Lord. Those who fight evil find that evil stays for the gloom of evil is everywhere. Christians who fight evil are harassed by evil on every side.

The Christian game is a strange one of faith. You remember the story of Peter walking on the water. There was a great storm. The disciples were in a small boat and were terrified of the humongous waves. Any moment, a huge wave could catch the boat and fill it. If that happened, all would seem lost. During the storm, Jesus walked by on the water. The disciples were extremely terrified then for they thought that they were seeing a ghost. When Peter realized it was Jesus, he asked permission to walk on the water too. Jesus said, "Come!" Peter came. All was good as long as Peter watched Jesus. Somewhere between the boat and Jesus, Peter saw the rushing waves. Once he looked at the waves, his feet sank. Peter was then being tossed by the waves until Jesus grabbed his hand.

Mom and Dad never focused on evil. They knew the price. Yes, evil is terrifying, but God is far more terrifying than evil. When we focus on evil, God lets us slip down into it enough to be terrified. Those who keep their eyes on Jesus experience the sloshing of love and the thrill of the call.

> 29 He said, "Come!" Peter stepped down from the boat and walked on the waters to come to Jesus. 30 But when he saw that the wind was strong, he was afraid, and beginning to sink, he cried out, saying, "Lord, save me!"
>
> -Matthew 14:29, 30 (WEB)

The caretaker died twice held. The caretaker took the faith of our dad and made it his own. The caretaker decided to be just like dad. The caretaker died in dad's arms. They were both held by their Savior.

> 9 This I pray, that your love may abound yet more and more in knowledge and all discernment, 10 so that

you may approve the things that are excellent, that you may be sincere and without offense to the day of Christ,

-Philippians 1:9, 10 (WEB)

JIM H DARNELL JR

Chapter 19

Visitors & Finding the Splash Zones

Visits from the United States

Mom and dad spent a lot of time hosting visitors from the United States. There was nothing they could contribute to the ministry. Mom cooked for them and took care of all their needs. Dad would take them around and show them the ministry. Dad was always so eager to pass on his vision of God's work. If dad ever saw it as bragging, he would have quit showing the work to the visitors. Mom and Dad just cared about people.

Orange You Glad You Got Hair?

Orange You Glad You Got Hair? I remember, one time, dad taking them on dirt roads to some villages. The heat was so unbearably hot that the air conditioner did not cool the car, so we rode with the windows down. One lady who visited us had silver hair. I was fascinated with how she could poof out her hair in such an orderly way. This was before perms were common. It seemed like every strand of hair was ever so close to the next one, but they never touched. It is strange what things fascinate a boy. Yes, I am still wondering about the value of this story because I have no point to go with it. Still, the tale might not leave some of you alone; it just might keep jumping in your mind when you are trying to kick it out. Anyway, I thought it was so cool.

Picking the story back up. We were driving down this dirt road. She had the most silver hair. The road was kicking up orange dust. I was an elementary boy, who paid no attention to things like the color of my hair, but I was fascinated by hers. After a while, I had to look at her silver hair again. This lady no longer had silver hair. Her hair had changed colors to bright orange, and her skin was still normally colored. I thought the silver hair was cool, but the orange was much more impressive. I don't know if she realized that her hair had changed colors or not.

31 Gray hair is a crown of glory. It is attained by a life of righteousness.

-Proverbs 16:31 (WEB)

Lost

This story has a way of striking at the deepest fears. An American pastor came to see us in Ivory Coast. He was traveling from Nigeria and had a two-day layover in Ivory Coast. He desired to experience the country, but his experience terrified him for many years.

The pastor's plane landed in Abidjan. He walked off the plane expecting dad to be waiting for him with open arms. This man got off the plane and saw no one there to greet him warmly. So, this man waited for 2 hours for dad, but dad never came. There was a reason that dad never came; dad did not know the man was coming.

The pastor figured that everyone knew my dad. The airport was an hour away across the town of nearly 1.5 million people. In his mind, all he had to do was ask, and everyone would point the way to dad. This man faced a few problems in communicating with the people. One is the pastor spoke only English and the official language of Ivory Coast is French. The pastor found very few people who spoke English. He kept asking and the people did not understand.

There was a second problem in communicating. The Africans had a tough time saying the letter "r" in French. To make it easier to say our last name, many Africans called dad, "Pasteur Daniel" Even if this man from the United States could have communicated with the Ivoirians, the name change may have been too much.

Phones were common in those days, the man could have phoned dad, but there was a problem. The man did not think he would need dad's phone number, so he did not bring it. If the man had brought the phone number, he would have been able to call from a phone booth. However, looking up dad in the phone book was impossible. The phone number was listed under Mission Baptiste (Baptist Mission). Without bringing the phone number, it was impossible to call dad.

Everyone had addresses; mom and dad were no different. All he had to do was get a taxi to my parents' house. However, like the phone number, he did not bring an address. Without an address, finding our home was impossible.

When the sun was starting to set this man started to get worried. All he had to do was ask for the American Embassy. It is their job to take care of Americans in distress. Such a quick ending to the crisis was not a part of this man's destiny for much worse was still to come.

He decided that he would get a taxi. He went to the taxi center and asked for an English-speaking taxi driver. They found him one. He got in, but the cab driver only increased the adventure. The pastor asked, "Do you know James Darnell?" The answer came, "No!"

So far, we have mentioned several things this man has done wrong. He did not tell dad he was coming. He did not bring an address or phone number. He did not try to contact the American Embassy. Still, there is more. He forgot to bring trip money.

This man got into the taxi with very little money. At this point, the man's main concern was no longer contacting dad but surviving. He told the taxi driver, "Go somewhere cheap!" The cab driver knew the rules Africans self-impose upon themselves. It was commonly understood that African life was too harsh for whites. Kindly, Africans spared whites the harshness of their lives because the belief was that African life would kill whites. This story proves that the African beliefs were not too far from the truth.

By African logic, the taxi driver would have taken the man to a hotel that catered to whites. If he had, the hotel would have figured out how to resolve this man's problems. The cab driver honored in part the stranded man's request. The driver took him to somewhere cheap, but it was 45 minutes across town. Such a long trip would have been expensive.

The hotel was in the middle of a black neighborhood of 200,000. Can you imagine this poor man being driven seemingly forever and wondering how much it was going to cost? After a while, the major roads were left behind, and they wandered deep into a community where the way out was lost.

The taxi driver pulled up to a cheap hotel. Dad had pondered upon this story many times. All he could ever figure out was that the taxi driver drove to a pocket in this huge neighborhood that spoke English.

The traveler paid the driver and got out. The hope was that on the other side of town, people would know who dad was. The man asked the hotel clerk if he knew Pastor James Darnell. The answer was still the same, "No!"

The clerk showed the man his room. The room was quite different from hotels in the United States. The room was fitted with five things. There was a bed that bowed like a hammock. It was obviously all worn out.

Over the bed was a mosquito net. The net was thick enough to keep out the mosquitos, but it also had holes to let the air pass. I don't know if you have ever slept under a mosquito net. Mosquito nets are extremely hot. They do not let air pass for the tiny holes only tease. Had one of the five elements been a fan, then some air could have been pushed through the net. In the tropics with extreme heat, humidity, and a lack of breeze, mosquito nets are miserable.

So far, we have covered two of the five elements: bed and mosquito net. In the corner of the room, there was a stick and a trash can. Dangling across the room was a rope attached to both walls. The clerk did not bother to explain beforehand about the last three items. The man asked, "What is the rope for?" The answer came back, "Oh that is easy. When the rats cross the rope, they go very slowly in the middle. You take the stick and hit them with it, then you put the rats in the trashcan."

This answer horrified the guest. He crawled under the mosquito net and stayed there all night. The bed was uncomfortable, and the net blazing hot, but that was the only place of safety.

In the morning, the man paid the clerk with the last of the money. A new problem arose. With a lack of money, a taxi was impossible. There was no way to buy food or something to drink. He could neither stay in the motel any longer nor return to the airport. Problems were killing hope.

In the morning, the man paid the clerk with the last of the money. A new problem arose. With a lack of money, a taxi was impossible. There was no way to buy food or something to drink. He could neither stay in the motel any longer nor return to the airport. Problems were killing hope.

The man picked up his suitcase(s) and walked out the door. He looked down one way of the street and then the other.

They both looked the same, so he chose a direction. He went to the end of the block and was faced with a new choice. He could either go left, right, or straight. Off to the side a block away, one of the signs caught his attention. It read, "Baptist Church!" (Likely, it was one of the Yoruba churches that dad helped organize.)

The man eagerly walked up to the church and entered. He found the local pastor and asked, "Do you know Pastor James Darnell?" The answer came, "I sure do! He is a good friend of mine! Let me call him for you." The local pastor picked up the receiver and used the rotary dial to call dad. Fortunately, not only was the local pastor at the church but so was dad at his house.

The local pastor greeted dad and told him that an American wanted to talk to him. The man quickly told dad his story. Dad responded, "I will be there in fifteen minutes." Sure enough, before long dad pulled up.

17 For though the fig tree doesn't flourish, nor fruit be in the vines; the labor of the olive fails, the fields yield no food; the flocks are cut off from the fold, and there is no herd in the stalls: 18 yet I will rejoice in Yahweh. I will be joyful in the God of my salvation! 19 Yahweh, the Lord, is my strength. He makes my feet like deer's feet and enables me to go in high places.

—Habakkuk 3:17-20 (WEB)

Two Reasons to Visit Africa

Usually, when people came to visit us from the United States, they came to visit us for one of two reasons. Some wanted to see God work. They were enthralled with the moving of God on the field. When they left, they all seemed satisfied with their visit.

There was a second reason why some people traveled halfway around the world. It was a compound reason. They were so terrified of the worst that could happen, and they had to prove that they were not afraid.

Normally, when people came, they started their talk with pleasantries like, "How are you?" "How long have you lived in Africa?" "Why did you decide to be a missionary?" These terrified people were too preoccupied for much of a greeting. It seemed like it was always the same. When we went with dad to pick them up at the airport, dad would greet them, "How are you?" "Did you have a safe flight?" Sometimes they would get out a "Fine!" Most of the time, dad's questions were never answered. The response was "Show us your worst."

Dad never liked to show off the worst within the country. He would ignore the statement and continue with small talk. Dad would continue with something like, "Did you have a safe flight?" "Did you get all your luggage?" The response was always the same, "Show us your worst." The other part that I never could figure out was why they always used the same phrase.

16 then it will happen that the sword, which you fear, will overtake you there in the land of Egypt; and the famine, about which you are afraid, will follow close behind you there in Egypt; and you will die there.

- Jeremiah 42:16 (WEB)

Encampment

Dad would eventually honor their request. He would take them there, but it never turned out well. The place was called

Encampment. There was this tiny area of land that no one had built upon. The poorest of the poor squatted there. There were tiny shacks one against the other. They seemed to all be built randomly. Trails were leading through there zigged and zagged. It was easy to get lost for it was just like a maze. I write about the stuff that amazed me. The place did not look like that terrible of a place to live compared to what the worst should be but looks are deceiving.

There were a few inconveniences like no running water or privacy. They could either wash up at the well or they could wash up in their homes after they had hauled the water. Either way, there was no privacy. The homes had many gaps in the walls. There might be 10 people living in a room sleeping one by the other. Either way, lots of people were going to see it. Everyone had the same problem, so it was usually not a big deal, but it was a huge price to pay for being poor.

I could continue with the story, but you need to think it through a little to understand why it was the worst. The Encampment had only narrow passages, huts, and nothing else. That was it. There was no running water – no porcelain. No one had sinks or toilets. So where do you go? I do not mean, where do you go to the movies? I mean where do you go when you have a business to attend to? Here is a good one, try unloading on the busy path that everyone must take. The litter would make everyone mad. They would settle the issue before the business was finished. So where do you go? We are not talking 10 people; we are talking about 250,000 people all with the same problem. Even the poor who eat little, still have to go. Even if there is an open area (which there wasn't), how could it accommodate the droppings of 250,000 people? Figure it out if you can! There is a reason why this place was the worst of the worst, and we have only started with the story.

I have wondered about this for many years. I understand that poverty is sometimes a treasure. The floors were made of dirt.

The floors were great places to bury the stuff that falls out. If it is buried deep enough, then the odor would not be such a problem. Once again, it is still not that simple. We are talking about a little bit of floor for a lot of people. How do you organize the holes so that the stuff has time to decompose? By the way, whose mat is going to be over the holes? The Africans usually slept directly on the ground with a thin bamboo mat under them.

The lumps may not be the biggest human waste problem. The flow leaves the ground soggy and stinky. How can anyone sleep with the smell of yellow water jumping up the nose? Unfortunately, this is still the good part of the story.

17 But I am poor and needy. May the Lord think about me. You are my help and my deliverer. Don't delay, my God.

-Psalms 40:17 (WEB)

The Two Seasons

It is commercial time! Ivory Coast had two seasons: the wet and the dry. The dry season lasted about nine months. Starting about June 1, it would rain for six weeks. At the end of the six weeks, the rain would stop for about three weeks. The rain would start again and continue for another three weeks which completed the yearly cycle. Another nine months might pass without it raining much again. There was either too much water or not enough water.

35 Yahweh, who gives the sun for a light by day, and the ordinances of the moon and of the stars for a light by night, who stirs up the sea, so that its waves roar; Yahweh of Armies is his name, says:

- Jeremiah 31:35 (WEB)

Flooding

Did you ever wonder why there was a patch of land that no one wanted to build upon? The only land available to the squatters was the flood zone. In some parts of Africa, the fishermen built their houses on stilts. During the floods, their houses would be above water. For them, living above water is not a problem.

In the Encampment, the houses were not built on stilts. The houses were built on the land. The one time I remember dad not just driving by, but taking me in, he told me a story. Dad said, "During the raining season, it flooded four feet." That was a very short story. I asked more, "Where do they sleep?" I knew that no one could sleep in water four feet deep, and the Africans were still there. He responded, "They slept in hammocks." The shack's posts must have been thick and sunk very deep to withstand the weight and the loosening of the shafts due to the water. I was quite concerned. I was not yet 4 feet tall. I would not have been able to stand anywhere. What did the kids do? The Africans rarely learned to swim for waters were always filled with deadly danger. I thought I finally understood why this place was the worst of the worst until I remembered, where do they go? It took me a while to get brave enough to ask my dad, "Where do they go?" Dad's answer was a little longer in coming, but he was straight to the point, "They go in the water, and it just floats. Many died!"

9 For you know the grace of our Lord Jesus Christ, that though he was rich, yet for your sakes he became poor, that you through his poverty might become rich. 10 I give a judgment in this: for this is expedient for you who were the first to start a year ago, not only to do but also to be willing.

-2 Corinthians 8:9, 10 (WEB)

Why Did They Need to See the Worst of the Worst?

I never made a friend with anyone who needed to see the worst of the worst. When they came, their minds were too preoccupied with what they longed to see. Once they saw their fears, they were still too overwhelmed to communicate. They never said that they were afraid. They did have a third favorite phrase, "I am not afraid. I am not afraid. I am not afraid." Many would chant this phrase over and over.

Afraid was not the correct word; they were terrified. I know what they said, and they meant it. They wanted to be brave and face their fears. They had to prove they were brave. I hope you understand that this story is not being told to be judgmental. The story is about them, and it is about us.

God can put His finger anytime He wants upon one of our fears. Right before the cross, Peter thought he was brave and would die with Jesus. Fear quickly consumed Peter. God can stir any strength into a paralyzing fear without notice. God is in the business of strumming fears. God's guilt and conviction are so powerful.

The proof is in the size of the test. If I ask you to jump over a marble, no matter how big your hops are, you cannot prove much. If I ask you to jump from one rooftop to another like in the movies, that is a different story. Some humans can make huge leaps and are impressive. No one jumps from one building to another when there is a major street resting in the middle.

Everyone wants to be proven great. The need to prove greatness never stops yacking. The voice inside that constantly boasts or condemns, "I have to prove my worth!"

People have to find reasons why they think they are worth something.

When God's tests start, He puts His finger on a weakness. The problem will consume until it is resolved, yet the problem has no answer. God's test is impossible to complete with human strength. Completing God's challenge is impossible even for the strongest humans. If humans can solve the test without God, it is not a very big test.

God places God size problems in our paths. The plan is for a godly answer to God's test. Those who refuse God's help harden their hearts and have invited in failure, guilt, shame, envy, lust, a bigger sense of being a failure, and to live every moment with God's disapproval. Those who trust God would fail miserably if God did not send miracles. To survive and thrive in the test, the believer must find the splash zone of God's love. There is no other way to thrive in God's test or even pass it without heaven opening up and filling the soul.

> 20 But, Yahweh of Armies, who judges righteously, who tests the heart and the mind, I will see your vengeance on them; for to you I have revealed my cause.
>
> - Jeremiah 11:20 (WEB)

What Happened to the Worst of the Worst Seekers?

The human answer is, "I don't know!" Mom and dad never shared with me stories about people returning home and the fallout of having seen the worst of the worst. I would imagine, there were many nightmares. We parted ways so I don't know.

> 6 being confident of this very thing, that he who began a good work in you will complete it until the day of Jesus Christ.
>
> -Philippians 1:6 (WEB)

Transformed by the Renewing of your Minds

Look at mom and dad. If we look at the beginning of the test, they did not offer much value. Dad was bullied by the outcasts. Mom's opinion did not count. She was very poor. She had great difficulty finding a church that would meet her needs and give her answers. She only went to one meeting on missions. Mom and dad would have greatly failed the test without God's intervention.

Look at the end of the test. Dad was the hero to many. Countless chose, "I am going to be just like him even to the point of dying!" God turned dad from being bullied to the hero to so many helpless people. Mom's finest hours were spent establishing Sunday Schools for extremely poor people. The one who could not find a church home much of her growing up years, gave a church home to many. It is in Sunday School that people dig in and try to find answers to life's problems.

Mom and dad had to have more than just the strength to do what they did. The Africans saw dad as the kind of man they wanted to be like. Dad was not recognized as the bullied one who either had to fight or take flight. Dad was a different person. Mom somehow went from needing answers to having answers. She was able to teach not only answers but teach teachers. Slowly God changed mom and dad from having nothing to becoming spiritual giants.

God's test has a purpose. God does not want to just give strength to overcome problems, God is in the business of changing lives. Scripture calls the maturing change, "transformation" or the "renewing of the mind." It is like when a caterpillar becomes a butterfly. One does not resemble the other. God changes broken and useless into the most beautiful things.

Let's go back to mom and dad again. God took their biggest weaknesses and turned them into their greatest strengths. We have already mentioned that God's tests require a God size answer. The bigger the impossibility, the greater the miracle needed to overcome. Let's take this argument to the extreme. If God wishes to give the biggest possible transformation miracle He could give, what must happen? What would the beginning and ending points be? Here is the answer. God would have to transform the biggest weakness into the strongest strength.

Transformation and the heavenly splash zone have something in common. Transformation comes in steps. After each step, heaven splits open and indescribable love, joy, and peace flood the soul. The overwhelming thrill of heaven always comes as God renews the mind.

Transformation or the renewing of the mind occurs when God changes a believer's outlook. A new understanding has taken hold and life is never the same. God's tests are too big to solve in human strength. God loves to take our biggest pains and weaknesses and transform them into tools usable for the kingdom.

9 He has said to me, "My grace is sufficient for you, for my power is made perfect in weakness." Most gladly therefore I will rather glory in my weaknesses, that the power of Christ may rest on me. 10 Therefore I take pleasure in weaknesses, in injuries, in necessities, in persecutions, and in distresses, for Christ's sake. For when I am weak, then am I strong.

-2 Corinthians 12: 9, 10 (WEB)

How does Transformation Happen?

Transformation occurs when the human way of thinking is replaced with the heavenly way of thinking. Human thoughts are replaced with godly thoughts. When God opens His mouth and speaks about the mysteries of heaven then life is forever changed. God speaks His love and His calling; something amazing happens. God must speak to make the old outdated. These words are words of life, so the heart and the mind are renewed.

If therefore there is any exhortation in Christ, if any consolation of love, if any fellowship of the Spirit, if any tender mercies and compassion, 2 make my joy full by being like-minded, having the same love, being of one accord, of one mind;

-Philippians 2:1, 2 (WEB)

How do we get God to Speak?

God does not like to argue. If He has to argue to make His point, He does not say anything transforming. If we do not want to listen to what He has to say, He just keeps turning up the guilt. Guilt is God talking disapproval. Why should God argue when He controls the guilt thermometer? God has the time to let us feel like our lives wither away to nothing. Hearing greatly improves when the lost-it-all feeling rules. When someone is ready to listen, God speaks. He always does.

The same thing can be said another way. The key to hearing is being still before God. As long as arguments jump into the mind, the mind is not very still. The mind must be unconditionally ready to receive whatever God has to say. God does not speak but guilt as long as there are ands, buts, or ifs.

11 He said, "Go out, and stand on the mountain before Yahweh." Behold, Yahweh passed by, and a great and strong wind tore the mountains, and broke in pieces the rocks before Yahweh; but Yahweh was not

in the wind. After the wind there was an earthquake;
but Yahweh was not in the earthquake. 12 After the
earthquake a fire passed; but Yahweh was not in the
fire. After the fire, there was a still small voice.

-1 Kings 19:11, 12 (WEB)

How do we Still the Mind?

There is one way and one way only to still the mind. Every argument must be surrendered to God. Surrender is saying to God, "You can have this, this, and this." Total surrender is saying to the Lord, "You can have it all! I hold back nothing!" Surrender to God is throwing out selfish thoughts to make room for godly thoughts.

10 In God, I will praise his word. In
Yahweh, I will praise his word.

-Psalms 56:10 (WEB)

How do we Stay in the Splash Zones of Love, Joy, and Peace?

Splash zones are incredible for a little bit, then they fade. The problem many Christians face is that they do not experience enough splash zones to keep life thrilling. Many believers have incredible life highlights, but most of life is shriveled up and barely bearable. The crisis these Christians face is that their lives have too few splash zones.

The key to a rich life is being good at collecting splash zones. None of the splash zones last very long, but enough short ones can make life feel extremely rewarding. The one who has a rich life is great at collecting splash zones. In theory, this sounds good, but there is a problem. We cannot make God bless us. The good news is that God has given guaranteed markers on where to find these splash zones. Anyone can go up to a marker and collect a splash zone. Splash zones are always found at the end of each transformation step. The only way to find splash zones is to be transformed. Transformation happens when God speaks into the vacuum caused by releasing selfish desires.

Transformation starts with surrender and ends with God speaking incredible things. There is only one step on the human side to stay in the splash zone of God's love – surrender. With there being only one step, the splash zone collector is constantly doing one thing – surrendering. To collect as much favor from God as possible, immediate surrender happens. Anyone who wants to live life with abundant love, joy, and peace eagerly surrenders to God.

The genius does not focus on what is perceived to be lost, but on what is gained. The perceived loss is the treasured dreams to prove love and greatness. The only loss of surrendering the will is a withered-up life; the gain of surrender is a life stuffed with the best riches of heaven.

8 Yes most certainly, and I count all things to be a loss for the excellency of the knowledge of Christ Jesus, my Lord, for whom I suffered the loss of all things, and count them nothing but refuse, that I may gain Christ

-Philippians 3:8 (WEB)

Their Visit to the Worst of the Worst

Dad was always willing to take our visitors from the United States into the Encampment. We would be sitting in the car in front of the Encampment, and dad would quickly explain to our visitors the tragedies of the place. Within just a couple of minutes, the answer was always the same, "I want to go back! I want to go back!" The sight from the road with the story was too much.

My Last Visit to the Encampment

It was just he and I. I always loved our field trips together. He took me in. I found out that the people were friendly and kind. They were all just hoping for an opportunity in life. Living there was their best hope of having a life, so they stayed.

We walked the crooked path. The path was not a path. It was all worn out. During the rainy season, the path would become muddy. The walking dug the path fairly deeply into the earth. The path was more like a bowl. At times, it was hard not to slip. Adults with their long legs could bounce from edge to edge of the path. With my short legs, I had to do a lot of walking in the middle of the path. The middle was slick and often slanted upwards or downwards. I fell many times.

Dad was compassionately moved by these people. Dad had done the only thing he could do for these people. Dad had started a church. The mission had bought a little building for $7,000 right in the middle of the place. Even the Encampment was an expensive place to live. Dad needed to give something to the pastor. The pastor was so happy to see dad that he kept smiling and laughing. Dad so loved the pastor and the community that he too kept smiling and laughing. Dad grew up as the poorest of the poor. Dad felt very much at home. It was one of my proudest moments with dad; dad saw the worst and the best came out of him.

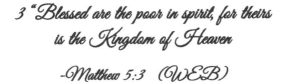

3 "Blessed are the poor in spirit, for theirs
is the Kingdom of Heaven

-Matthew 5:3 (WEB)

So, What Happened to the Worst of the Worst Seekers?

God is eager to play the transformation game so that He can speak the most thrilling things and give the richest gifts.

Giving away the treasures of the universe is God's plan for those who are ready to receive.

Imagine with me for a moment what kind of transformation leads from the worst of the worst to the best of the best. God takes a terror of failure, poverty, and hopelessness and turns it into ... Tell me!

There are other places with poor people full of a sense of failure and hopelessness with someone's name written all over it – someone who was afraid of the worst of the worst until God spoke. Now, being with the worst of the worst is the best of the best places to be. It takes fully understanding the worst to be able to give the best.

God's plan is to turn the worst of the worst seekers into givers of life. God does not want to diminish the fascination with the worst of the worst, just turn them into His ambassadors to the worst of the worst. However, God's plan is followed to varying degrees.

1 Blessed is he who considers the poor. Yahweh will deliver him in the day of evil.

-Psalms 41:1 (WEB)

Three Choices

Those who came to see the worst of the worst had three choices. The first choice was to dwell in fear; see and be more terrified. They could refuse to surrender at all to God and are therefore inviting the terrors to destroy their own lives. The only way they could shield themselves from their nightmares is to harden their hearts and become less sensitive. The first choice of totally refusing to surrender and rejecting transforming fully ruins life.

The second choice is to live a withered life. Surrender only happens when life is too withered to stand anymore. All they choose is a little mist rather than bathing in God's love. To a degree, the marginally surrendered are useful to the kingdom. They often fill churches and pulpits and consider themselves faithful, but they teach people how to keep a multitude of sins safe by trusting little.

Those who have chosen choice three dwell in the presence of God. They have chosen to become transformed at the maximum rate. They need minimal guilt. Upon experiencing guilt, they surrender immediately. As one splash of God's love fades another happens. The Giver of Life transforms the broken and ugly into the very picture of God Himself. God speaks to the surrendered regularly; it is like the two become one.

47 Therefore I tell you, her sins, which are many, are forgiven, for she loved much. But one to whom little is forgiven, loves little."

-Luke 7:47 (WEB)

Why does Surrendering sometimes not seem to Work?

Sometimes the process of surrendering does not eliminate the dried-up feeling and brings a renewing of life. Surrender happens, but nothing else happens. God's plan is always the same. God is not stingier with his treasures at some times than others. Even so, at times God chooses not to reward surrender.

If the rule always has been and forever will be the same, why does it sometimes not seem to work? The guilt and dryness come, so something is found to be surrendered, and God remains silent.

God will not be conned. When a father demands the keys to the car, and all he gets is the key to the trunk, the father does not think it is very funny.

God wants it all. If given a small part, the Father is not amused. The thrill and the blessing will not come until He receives the full surrender as prescribed by the guilt.

Sometimes something major is surrendered without trying to hold anything back. God neither blesses nor removes the guilt. God knows what all is even when we do not understand. The guilt drives on the search for more. Soon enough God's finger can be felt upon more sin. Once God is satisfied that all He put His finger upon is surrendered, God's pleasure comes powerfully.

7 Don't be deceived. God is not mocked, for whatever a man sows, that he will also reap.

—Galatians 6:7 (WEB)

JIM H DARNELL JR

Chapter 20

Innovative Ways of Reaching the Lost

Correspondence School

Since Christianity was so different than traditional African religions, dad was very concerned that the new converts did not understand the decision they made. Even if they did understand, they needed a foundation that was often difficult to find. At new church plants, no one knew enough to explain anything to the others.

Mom and dad offered a 13-week correspondence school to all new believers. The course was very affordable even for the Africans. They charged something ridiculously small just so the participants could feel like they bought it instead of being given something free.

Dad initially wanted the course just so new converts would have the foundation of the basics of Christianity. However, the course became popular. People all over Ivory Coast were taking the class. It just seemed like random people, but they were taking the class. The class was too successful causing dad problems. Dad did not have the time to keep up with the numerous students and his other work.

Living near us were two single women missionaries: Estelle Freeland and Wilma Rodgers. We often saw each other because those ladies were very active in church work. One day, both of those ladies were outside, and dad cornered them. Dad said, "I teach this new Christian correspondence course. It is supposed to be for Christians, but many are taking this class all over the country and are getting saved. It is becoming too much work for me. Would one of you like to take it over?" There was a pause, and neither lady answered. Finally, Wilma answered, "I will do it!"

Wilma continued to teach the course. The course became more popular all the time. It was not too long until the course became full-time work for her.

Dad's vision of offering the course proved too much for him. Letting Wilma take it over should also be considered her work, for it was a ministry. Countless numbers of Africans otherwise unreachable found Jesus. Through the course, Wilma had one of the most amazing missionary careers. In a male-dominated society, this single woman had a huge impact on God's kingdom.

At the end of the course, the student would receive a diploma. Most of the students had marginal education. Some took the course to practice their French, others took the course for the diploma at the end. Probably most had never received any formal recognition for their accomplishments. Can you imagine the joy for an adult to finally receive a diploma? That diploma was so special that it would be shown to the whole family. It was common for the diploma to be the first in the family.

How long do you think it took for some to think, "I want a diploma, too!" Family members and friends who lived in the villages and other places far away from the church plant would enroll too.

Mom told the story of one location over 100 miles away that had a lot of students who had graduated. There were so many that dad went to see them. At first, there was a lot of tension in the air. After a little bit, the Spirit of God fell upon the group. It became very obvious that the course was very effective in both leading people to Christ and developing a Christian foundation.

13 Come now, you who say, "Today or tomorrow let's go into this city, and spend a year there, trade, and make a profit." 14 Whereas you don't know what your life will be like tomorrow. For what is your life? For you are a vapor that appears for a little time, and then vanishes away. 15 For you ought to say, "If the Lord wills, we will both live, and do this or that."

- James 4:13-15 (WEB)

Crowd Collectors

Dad had come up with several plans to draw crowds. Dad sometimes invited people from the United States who were good at drawing crowds. These activities were rarely associated with church starts; instead, the crowd-drawing activities were for the existing churches. When dad brought someone in, he tried to utilize the drawing power in as many churches as possible.

One of dad's favorites that never happened was Louis Armstrong. He was the strongest man in the world. Louis was so large, that he would have required two first-class seats. After studying the cost, dad chose to pass.

Dad did bring in a strong man. This man had several stunts. He would lay on a bed of nails. For another stunt, a huge rock was placed on his stomach while lying down. He asked for the biggest and strongest man there to come forward and try to break the rock with a sledgehammer while the stunt man was singing. When the singing ended, the stunt would be over. The local strong man was to quit slinging the sledgehammer upon the completion of the tune. This huge man was swinging the sledgehammer as hard as he could, and the man on the ground stopped singing. The stunt should have been over, but the local man continued to swing the hammer with all he had. Dad and a couple of other men rushed in to stop him. The Africans were very impressed.

Arthur Blessitt impressed the Africans. He has walked the world over carrying his cross. In his pilgrimage through Ivory Coast, Arthur touched many lives. They saw a man who loved Jesus and was willing to walk through every country in the world to share the Gospel. His passion touched many lives. Arthur was like a breath of fresh air. Everyone came face to face with the magnitude of the calling.

Dad called for the churches to have a weekend camp. Several thousand came. Twice as many came as paid. Those who could pay most likely paid. For many, this was a turning point. A lot of these churches were so isolated, that it was like they were the only believers around. When they saw so many other believers, strength was gained. A large number of others received salvation. The weekend ended with a renewed zeal to seek those who were not yet saved.

Dad always tried to figure out the best way to draw crowds. Crowds would often come in large numbers. Crowds were another manner dad chose to reach people for Christ.

34 Jesus came out, saw a great multitude, and he had compassion on them, because they were like sheep without a shepherd, and he began to teach them many things.

-Mark 6:34 (WEB)

Chapter 21

Times of Revival

No Revival Here

One of the church plants that I remember best, flopped. I recollect going around with dad and greeting people in a village. Several were interested, although the number was smaller than normal. The village was on the side of a hill. Dad always had to use whatever location he could find. There was a man who had a huge shade tree in his backyard. Dad asked if they could meet under the tree. The man graciously agreed.

This man loved the prestige of having the Americans come to his house every week. This man always attended the meeting, but he was never really interested in the Gospel. He did not reject it; he was just lukewarm. I imagine during the week, he lived with the same vices he had before. We soon came to realize that nothing was going to come of this church plant. After a few weeks, dad shook this man's hand and we parted forever.

I could tell that mom and dad were quite hurt about the failed attempt to start the church. Two other church plants failed. One thrived, but when dad turned it over to another missionary, it failed. The other started well, but we ran out of time before returning to the United States. We witnessed salvations, but there was no time to organize a church. Dad had to try anyway in the short amount of time we had.

Back to the church plant that failed in the backyard. Mom and dad realized that by default, the owner of the house always puts a ceiling on growth. The church's growth will not exceed the owner of the land. Mom and dad did not get angry, fuss, or gripe in any way. They just decided that they would start churches on neutral ground in the future.

11 Whoever will not receive you nor hear you, as you depart from there, shake off the dust that is under your feet for a testimony against them. Assuredly, I tell you, it will be more tolerable for Sodom and Gomorrah in the day of judgment than for that city."

-Mark 6:11 (WEB)

This neighborhood had a tragic beginning. The land was private, but many squatters came anyway. Those who were too poor to live in town built a little shack on the edge. This area became filled with people. The people were notified to leave for the owners were going to bulldoze the area. Those living there just stayed. They had nowhere else to go. Besides, the people had a hard time believing that the homes of so many would be crushed. On the appointed day, the bulldozers came. They just plowed into the flimsy buildings. Suddenly, the people started scrambling. What little they had was in the way of destruction. They grabbed what they could. The area was flattened. To make room for the buildings all the trees were torn down.

After the failed church attempt, dad knew that the next attempt needed to be in Yopougongare. Dad wanted to find a location in the middle of the neighborhood. Since the neighborhood was so jammed with houses, there was no empty land to build a church. Although dad did find a building for the church, dad wanted the first meeting to be out in the open for all to have an opportunity to choose. In the middle of the neighborhood, one tree had grown so massive that it was impractical to uproot. Under the tree was an open area. God had reserved for Himself a starting point for the church.

On Saturday, dad had visited the neighborhood. I wished that dad had taken me with him. It would have done me a lot of good to see God prepare the hearts for His word.

Sunday morning, Patty, Julia, and I packed a lot of chairs. We packed the car full of chairs while somehow leaving tiny spots for Julia and me in the back seat. I asked Dad, "Do you think anyone will come?" Dad answered, "People will come!"

When we arrived, no one was there. No one ever arrived before us for that was not cool. Once we arrived, people started coming from all over. I was concerned that we might not have enough chairs. The car could hold 43 chairs and that was it.

People quickly unloaded the chairs. Hymnals were placed on the seats. Everything went perfectly. We had never seen so many people on the first service. Every service, the church kept growing. Mom had her hands full trying to organize Sunday School. It was not uncommon for the other churches to grow large, but these numbers were extra-large.

All the land had been built upon. There was not any land to build a church building. Many houses were built with the same floor plan. There was a modest living room. Off the living room were tiny bedrooms that looked more like closets than anything else. Open to the living room was the kitchen. There was a bathroom, but it was not much either.

To save land, these houses had been built in a duplex form. The floor plans were the same except the opposite. The living rooms were built up against each other. This church start was the first-time dad had purchased a meeting place before the work began. Dad had acquired a whole duplex and had the center wall knocked out. The two living rooms became one room. The church had a place to meet. The duplex became the church, but it also was the pastor's house. Dion Robert was truly an amazing pastor.

Although there was a place for a new church, dad chose the open air for the first meeting to draw in the people. I felt like we were on stage. People all around were standing and observing. Now and then, one of the observers broke ranks and left the observers who stood at a distance. These brave souls came and stood beside us to sing to a foreign God and seek out His Maker.

Dad had other chores to do that were not so pastoral. Dad had to provide the church with churchy furniture. Dad had to buy the lumber and find people to build the furniture. Someone had to haul the lumber and pick up the benches. The issue was not just volunteers, vehicles were very scarce among church members. Dad had to haggle a lot to get the most affordable price.

A basic pulpit was built. The congregation needed benches. These benches were nothing more than an unpainted yet sturdy seating area. The benches did not have backs. This little inconvenience ended up being so useful.

There was a problem almost from the beginning. The building was too little. Dad could not secure more funds; the church was growing like crazy. There seemed to be no answer; yet it was at those times, that Dad found unique answers.

Dad called his missionary friends. Dad exhausted all his contacts for ideas. Someone told dad about cell churches. The cell movement had not yet become a rage. The cell movement was designed to get people to meet in homes as well as the church building. The concept was that churches could grow faster if at times they met in homes. The cell leaders would have to recruit church members and friends to make cells adequately large. The cells would also reunite for collective worship in the main auditorium.

The traditional cell model would not work for Yopougongare. However, a light bulb went off in dad's head and he was so excited. Dad's idea was to divide the church into cells. All the church members would belong to a cell. The cells would rotate meeting in the main building with the pastor. The problem kept growing. Cells became too large and kept dividing.

It was not too long afterward that cells became the fad in many places around the world. Traditional Sunday Schools were seen by many as outdated. Many would look upon the cell model and say, "See, this is why every church should abandon Sunday School and go to cells." Things need to be kept in perspective. The church from its infancy always grew fast even to this day. The cell structure did not create growth; the cell structure kept up with the growth. Many things were in place that generated the growth. Any church that ignores the other things that generated the growth and hails the cell structure as the promise for growth may be deceiving themselves. Cell structure combined with other things was the perfect combination of the day.

God has a sense of humor. Mom and dad's plan was to never to meet in the house again due to the ceiling imposed by the house owner. For the next church start, the genius and brand-new idea was to meet in houses. House meetings worked this time because the house owners were Godly. God replaced an old wound with wonderful memories. Not long ago, I mentioned to mom about the failed church start. She said, "I don't remember!" God loves taking wounds and turning them into something beautiful. If you are like me, I still have plenty of sorrows that God is not finished with yet. I long for the day in which God speaks new memories that makes the old irrelevant. Those days do not need to come, but God in His timing often sprouts joy from beds of sorrow. It is the sorrow that makes the new memories so wonderful.

The story about the church in Yopougongare is far from over. Six years passed before we said, Goodbye to Ivory Coast forever. Let's pick up some of those things before we finish this story. These other things will add clarity to why church growth was phenomenal in most of the churches.

8 Others fell into the good ground, and yielded fruit, growing up and increasing. Some produced thirty times, some sixty times, and some one hundred times as much."

-Mark 4:8 (WEB)

Why Add when We can Multiply

Math was always my best subject in school. Nothing else came close. The incredibleness of math is something to be celebrated. I know, some of you are about ready to puke at this moment.

Dad knew that there was no way, he could reach Ivory Coast by himself. Dad believed that he could generate all the churches he could, and the country's spiritual needs would remain relatively unmet. Dad was convinced that the churches themselves had to be passionate about starting work also.

5 So the assemblies were strengthened in the faith and increased in number daily.

-Acts 16:5 (WEB)

Oh, Pastor …

Most of the time, when someone came up to dad and said, "Oh Pastor," they needed money. Most were extremely poor. They just had nothing. None of their friends had anything. If someone had something, then countless people around them would ask for it. Don't get me wrong! The Africans were not wanting to be beggars; they were just desperate. Dad gave generously. Mom never complained. Mom always trusted that Dad followed God's leading.

21 He who despises his neighbor sins, but he who has pity on the poor is blessed.

-Proverbs 14:21 (WEB)

Bring them in

Dad never got the idea that the Africans needed time to mature before making mature decisions. In dad's mind, the moment after Salvation, they were to start walking like spiritual giants. Just because they did not know much was not an excuse to ignore the Great Commission by Jesus. The Africans did not know that they needed time to mature either. As a result, the Africans just kept bringing them in. The norm, not the exception, was to tell everyone they knew. Family, friends, and acquaintances were all told about Jesus. Everyone was telling about Jesus, and people just kept coming.

Therefore, leaving the teaching of the first principles of Christ, let's press on to perfection —not laying again a foundation of repentance from dead works, of faith toward God,

-Hebrews 6:1 (WEB)

Your Turn

Dad would often go back to the established churches. Dad would preach, "It's your turn!" A stranger had gone to them and given the Gospel, now it was their turn to reach out with the Gospel. These churches at first seemed a little shocked. They thought that they were doing the Great Commission! They had faced the prospect of losing everything when they told their families about Christ. Many forfeited everything for the sake of the Gospel. Not only did they tell their families, but they also told their friends as well. Many lost their friends as well as their families. It was to these highly evangelistic souls that dad challenged them to take their turn.

> 33 "But this is the covenant that I will make with the house of Israel after those days," says Yahweh: "I will put my law in their inward parts, and I will write it in their heart. I will be their God, and they shall be my people. 34 They will no longer each teach his neighbor, and every man teach his brother, saying, 'Know Yahweh;' for they will all know me, from their least to their greatest," says Yahweh: "for I will forgive their iniquity, and I will remember their sin no more."
>
> - Jeremiah 31:33, 34 (WEB)

A New Problem

Many had faithfully taken the challenge and told their families and friends. They kept telling their loved ones. After telling the immediate loved ones, they reached out to the others. They kept telling and kept telling regardless of the cost until they ran out of contacts. After a while, they had not one new soul to tell. They had run out of acquaintances, but the desire still burned. The passion that drove them to risk it all for God's sake did not die. The passion was as strong as ever, but the opportunities became scarce.

> 3 This will we do, if God permits.
>
> - Hebrews 6:3 (WEB)

Something New

A part of maturing is looking beyond personal problems and

seeing the needs of others. In the churches, the maturing was happening. One time when dad heard, "Oh pastor," the request was quite different. "Pastor, our hearts are heavy for our country. Will you take us out so that we can witness?" They just wanted to tell others about Christ. They had exhausted their contacts, but the fire was still burning.

> 3 He said to me, "Son of man, can these bones live?" I answered, "Lord Yahweh, you know." 4 Again he said to me, "Prophesy over these bones, and tell them, 'You dry bones, hear Yahweh's word. 5 The Lord Yahweh says to these bones: "Behold, I will cause breath to enter into you, and you will live. 6 I will lay sinews on you, and will bring up flesh on you, and cover you with skin, and put breath in you, and you will live. Then you will know that I am Yahweh."'"
>
> -Ezekiel 37:3-6 (WEB)

Their First Journey

On their first journey to random evangelization, dad went to the location to pick up the men. There were four others. These 5 went to a village just to talk about Jesus. Most Africans speak multiple languages. They found a village they knew nothing about. The village was only 10 minutes out of Abidjan. Between the five men, they spoke nine languages.

Eagerly, the men got out to share their faith. The first men they talked to could not understand. They all took turns jumping from one language to another. The villagers just kept shaking their heads. When they ran out of languages, they decided to find more men. The same thing happened. They went all the way through the village with the same result. They could not find anyone in the village that they could talk to. These five men were left with burdened souls. Who was going to tell these villagers about Jesus?

The crew moved on to another village. God was compassionate to the evangelizers, and some believed. Their hearts were rejoicing, but they were also so heavy. This inability to communicate with the one village fueled the flame within. This one little event of being unable to share the Gospel was about to have a worldwide impact. The change was bigger than they could ever have imagined. The change was from a longing to reach their world for Christ to a determination to reach their world for Christ.

46 Blessed is that servant whom his lord finds doing so when he comes. 47 Most certainly I tell you that he will set him over all that he has.

-Matthew 24:46-47 (WEB)

The Prophet Had Come

These journeys became highlights of their lives. Dad would talk more about these journeys than perhaps anything else. The most important things, dad only said once and briefly. The joy of these trips would not let dad be silent. Meeting one on one and sharing the Gospel blessed him so much.

As a child, I asked my dad to tell me more of the story so that I could retell it correctly. Dad could not. I would have pushed him, but I could see that he was unable to express more. Dad gave me something like two sentences which was twice as good as the one sentence he gave me at other times. It is not what dad said that mattered most. On that day, dad glowed more than I remember seeing any other day of our lives. The beauty was seeing how God so touched dad. I had to be satisfied that I got all the stories that God intended. In the retelling of the story, God is not permitting me to state only what I heard from my dad. In a way, I do not know the whole story, but then again, somethings are certain anyway. God's timing, conviction, Salvation, and joy are all quite predictable. For me, the challenge is – have I paid close enough attention to God's process that I can fill in the other parts of the story that only the spiritual eyes can see?

These soul seekers went to one village. The harvest was so ready. A stranger had already visited the village in times past. He came with a proclamation, "A white man will come someday and tell you about God!" For years, these villagers just wondered about these strange words. The Spirit of God is not idle. God had been working on the hearts of the villagers. A phrase like "The can is blue" is not that important, but the proclamation that God was going to come was important. God had been working and stirring the hearts about the most essential. God was asking a far more pressing question, "When I come through the white man, will you be ready?" God has so worked on the villagers that they went from concluding that God was going to come to needing God to come. God's movement was so powerful that the "someday" needed to be "today." The words of their prophet echoed so loudly within. The villagers realized their lives were on hold until the day of God's arrival. The time came when the words needed to be today.

A strange car pulled up. The villagers were not looking for an old friend who bought a car, they were looking for white skin. The villagers casually looked into the car to find white. The driver looked white; their interest was pricked. The car stopped, and the doors opened. A lot of men came crawling out. They were not white. The driver's door opened, and a white arm surfaced. Sure enough, a white man stood.

Dad and his fellow brothers walked into the village. The other men were Black, but there was one who was not. Amid the strangers was a white man. The village could not ignore that a white man had come. Dad could have been a doctor who had come with compassion, but he was not. The white man could have been a dentist willing to pull rotten teeth, but he was not. He could have been a tourist, or a businessman looking for the right area, but those were not the reason for the visit. Being a pastor was only one of the countless things dad could have been. The village may have had some sick or mouths with rotten teeth, but the village was only looking for one kind of white man. Every time a car came, the villagers noticed if there was a white man. The prophecy called for white, and a white man was sure to come. The village expected that a white man would come, and he would do one thing: tell them about the Living God.

The villagers could not come running or wait too long. The only conditions they knew for the one who was to open heaven was the person was white, and a male. A white man had come. He neither brought anything with him like dentistry equipment nor did he pull out his wallet to buy anything. The white man and his friends walked around greeting those who were there. All the villagers could think, "Is he the one? Is he the one who will tell us how to make peace with the Almighty?"

The kids ran around telling all adults who did not know, "The white man has come! The white man has come!" Even the little ones knew that a special white man was going to come from God. Before long, the village had assembled. They only wanted to know one thing. Is this white man from God or not? Is he the long-awaited? He had to be the one. The crisis has already come. They could no longer stand the wait – they needed a savior that day. The prophet's words had made their entire world revolve around one thing, the arrival of God through the white man.

Dad and his fellow men opened their mouths and said, "We have come to tell you about Jesus, the Son of God!" All the villagers said the same thing, "We know. We have been awaiting you! We had been told that a white man would come and tell us about God!" All the men of the witnessing party were stunned. God had already come with such power. Their whole bodies tingled with supernatural power. On that day there was an incredible collision. The anticipation that God would come slammed into the reality that God had come. The arriving party's burning passion that they had to tell someone about Jesus crunched into the appointed village. God's divine appointment had come.

The newly come started explaining to those around. The villagers did not need to be convinced; they just needed to understand. As the newcomers spoke, God revealed Himself. The harvest had come. The villagers opened their hearts to the Creator of all the universe. There was much joy upon the arrival of the appointed one, but a far greater joy was coming. So many of the villagers were soon to become children of the Highest.

All over the village voices echoed, "Dear God, forgive me for doing wrong. I invite You to come into my life. Thank you for forgiving me and coming into my life. I am willing to follow You to go and do whatever You ask, in Jesus' name, amen!"

After that prayer, God did what only God can do. The heavy layer of guilt that sucked the thrill out of life blossomed into an indescribable joy. The burden of hopelessness transformed into the most incredible wonder. Smiles attacked the faces. There was no question that they had met the One True God.

The words of the prophet had grown from curiosity to a heartbreaking need. When the village was ready to respond to God even before they understood, the time for God to appear had come. God's plan was so brilliant that everyone was overawed by Him.

We could save the word, "Prophet!" for the great men of oldwho foretold amazing things. God is still in the business of relaying messages about the future through appointed people. If God's messengers who foretell what God is going to do are not called prophets, then what should they be called?

Dad and his friends knew nothing of the words of the prophet who had once passed through. These evangelizers had picked a random village from their perspective, but God had ordained the moment. Every detail had been carefully weaved by the Master's hand. Two miracle timings had to occur at the same time for the village to respond as they did. The village appeared to be picked at random, but it had to be the village prepared by the prophet. The moment of arrival needed not to be a moment earlier or a moment later. When the people were unconditionally ready to respond to God, the evangelists came. The total randomness from the human perspective made God's handiwork so beautiful.

Measuring the Blessing

Who was blessed more? Was it the prophet? Were the villagers who received Christ? Was it dad who was prophesied to come? Was it the men with dad who asked dad to go and were partakers in the harvest? Who was blessed more? The question is important, but a little misleading.

Let's start with the prophet. He did not lead the village to Christ. He did not see the outcome, but he did. God had told him in advance that a white man was going to tell them about God. It was not the prophet's job for his responsibilities were to foretell a miracle to come. If he was to foretell a miracle, then he knew the miracle was coming. Imagine the time after he told the village. God reaffirmed in his heart, "The harvest will come because you told them. Had it not been for your faithfulness, the harvest would not come. You don't see it with your eyes, but the harvest is yours! I am so proud of you, you did exactly as you were asked." The prophet may have been a very ordinary man who was not in the habit of prophesying. What he saw with his spiritual eyes and reaped in advance may have been the highlight of his life.

The villagers received salvation. Eternity with God is the biggest of gifts. The villagers were so convinced in advance that God was coming, and God did come. They were burdened by a heavy load of sin and guilt. Their rebellion toward God had wrecked their lives. God generously gave salvation. The guilt, shame, and heaviness transformed into the sweetest experience of their lives up to that point.

The men who went with dad reaped greatly as well. They appeared on the surface as indirectly associated with the prophecy. The prophecy was about a prophet, a village, and a white man without a fourth party listed. They and dad had gone as one. Their bond from previous evangelical trips united them as one in the Spirit. When they heard the prophecy, the Spirit of the Living God reaffirmed that it took the whole group to share the Gospel with the village. The group talked to people. God made them His hands and feet. Each man of the group participated in His God-given portion of the harvest. God dumped His love on each of them like He was so completely proud of them. The cool part of it being a group is that each one was utterly amazed by God, and they had each other to share it with. Each of these men went from being completely insignificant in a brief period of time, to God's ambassadors to a lost world.

Dad! I could talk about dad, but how could I say that his blessing was more important than the others? When dad came home, he was so excited that he could hardly talk. This was the most excited I saw dad in Africa. The time had come for God to bless them all well beyond any human measurement could tally.

The randomness of life only appears haphazard. God organizes life together where the beauty of his planning will amaze for eternity. I limited the discussion to those involved in the prophecy. Still, God may be talking to you and saying, "I love you as much as the prophet, the villagers, the men who shared the Gospel, and the white man. My love for you is that real and I am just as excited about my plans for you as I was in sending the prophet." Tell me if you can, who is blessed more?

14 I will satiate the soul of the priests with fatness, and my people will be satisfied with my goodness," says Yahweh.

- Jeremiah 31:14 (WEB)

Chapter 22

Yopougongare Conclusion

Contagious

Let's briefly repeat the story about the business meetings. Dad would tell the main church leaders what needed to happen to find the best solution. The answer was always the same, "Oh Pastor that is a good idea, but you do not understand for that will never work here." So, they rejected dad's plan. Once the business meeting started dad said little except approval for the good parts and ask questions. After enough approval and questions, the church members brilliantly came up with the very same idea that they had rejected in the first place.

This story had to be retold because you know where it is going. The church members had a wonderful idea, it was their job to reach their world for Christ. Once that decision had been made to reach their world for Christ, the Africans were on a mission. Dad's strange words that they need to be on missions to save Ivory Coast came to life as their very own.

> *3 He will be like a tree planted by the streams of water, that produces its fruit in its season, whose leaf also does not wither. Whatever he does shall prosper. 4 The wicked are not so, but are like the chaff which the wind drives away.*
>
> *-Psalms 1:3, 4 (WEB)*

No Obligation

The issue was not a sense of obligation; the issue was being compelled. Obligation stems from a moral responsibility to do right. They were compelled to witness. God's love so moved within them that all they could think about was reaching others. The cycle never ended. They surrendered their will to God. God filled them with love and a passion for others. To stay in that love and passion, they had to open their hearts even more to God by surrendering to new frontiers of faith.

11 Now that no man is justified by the law before God is evident, for, "The righteous will live by faith.

-Galatians 3:11 (WEB)

Yopougongare Six Years Later

Six years later, Yopougongare's main church building was still two little living rooms that were connected. Church attendance was 600 people. The church was divided into nine cells. Every Sunday, two cells would join together in the main auditorium for service. The other seven cells would meet in their respective homes. They only had the chance to meet in the main auditorium about once a month.

46 Yahweh lives! Blessed be my rock. Exalted be the God of my salvation,

-Psalms 18:46 (WEB)

Our Final Service at Yopougongare

I do not know how they chose who could come to the final service. Not being able to attend must have been difficult, for they knew this was likely our last time together. The Africans paid incredible prices to go to church. These traditions that are described were commonplace in many churches. It does need to be noted that these customs set forth were perfected to the point that no one needed to be told what to do.

Everyone greeted us warmly. We shook hundreds of hands. The eagerness to shake our hands was overwhelming. Even as a 16-year-old athletic boy, they crushed my hand. For them, we were one. We did not save; God did. They chose to be just like the God they saw in us. I am very thankful that God presented a picture of Christianity far greater than we were, but we were their heroes of the faith. Mom, dad, Patty, Julia, and I were faithful to our calling. The Africans were so extremely grateful that someone had brought them salvation.

Once the service was about to start, we were ushered to a pew. Men sat on one side, and the women sat on the other. I thought, "This was like Puritan days!" Men were allowed to sit by women, but it was not practical in churches like this one. Space was of the essence.

Men and women are built differently. I know that I can be getting myself in a lot of trouble, but I am just trying to tell a story. No matter how big or small a woman is, the hips usually are at least as wide as the shoulders.

Fit men have very different shapes. The men did physical labor for a living. Many worked in construction. Very little was electric or diesel-powered because human labor was so cheap. A lot of men made bricks for a living. When the building under construction had more than one floor, they would just throw things up to the next level. When an auto mechanic needed to move a car engine, he would just grab the motor and lift it. These African men were huge, but not in all places. In comparison, the men's hips were tiny and shoulders extra wide.

Even under the cell system, the church tried to fit in as many as possible. Women sat on one side with hips arms shoulders touching. This made things extremely hot for the women. With the extreme heat of a typical tropical day, bodies touching just magnified the feeling of hotness.

I just marveled at the men. The men sat hip to hip. A lot of men could be squeezed into one row this way, but there was the shoulder problem. For many of these men, the shoulders were twice as wide as their hips; this was a major problem. These were massive men. Every man would tilt his right shoulder back. The men made some sort of Dagwood sandwich. The man to the right would place his shoulder on the man's chest to the left. The man to the left, his shoulder would be pressed against the man to the right's back. It was quite common for every man's shoulder to stretch from one man's backbone to the next man's sternum. Some men were so large that their shoulders stretched past the backbone and sternum. The women were very hot. I cannot imagine how hot those men were. Still, they made their choice.

I was mesmerized by the sight. No wonder the men could not sit by the women. All the pews were filled to maximum capacity except one. Every pew was filled with black people except the half-empty one. Some could have sat on the end, but no one did. They had too much respect to fill that pew. One pew had four white people: mom, Patty, Julia, and me. I don't know about the rest of the family, but I was thankful for the generosity. I was highly embarrassed that I had so much room and they had so little.

The center aisle was kept open. The pews on the right side facing the front were up against the closet-sized bedrooms. These bedrooms were all open. As many chairs as possible were placed in the bedrooms.

The space behind the last pew was standing room only. All the women were seated, and the back was filled with men pressing tightly one against the other. Every available space within the building was occupied by people pressed together.

The Africans did not have air-conditioning. The buildings had huge doors and large windows to allow as much of breeze to flow through as possible. All the doors and windows were opened. The crowd pressed tightly and continued out the door. The crowd continued as far as the amplifier could reach. The conditions were the same for the windows. People stood near the windows, and they were stacked pressed one against the other as far as the sound traveled.

They were so well organized; it was evident that this was normal. I never once heard complaining. I saw a lot of sweat and fanning. It was a hot day. By noon it was probably 100F and 90+ humidity. I felt myself steaming and I touched no one. We were also given a pew under the fan. Many people were very sweaty even before they interlocked into their positions at the beginning, for they had to walk in the sun to arrive.

Most of my memories of African days are gone, but this day has forever echoed in my mind. I thought that I had seen the most shocking parts, but the best was still to come. The room was filled with men and women of legendary proportions, and it would take many decades for me to appreciate the incredible company I was in.

The song service was a typical one. Most people sang very loud. Many sang off-key, but no one cared. They had two things going for them. Everyone knew all the songs by heart, so they sang louder because they did not have to think about the words or the music. The second issue is that all of them had risked their lives for Jesus; there was no turning back. Singing to God was like singing in triumph after a major war victory. They sang with great enthusiasm.

Dad preached his vision for their future. He challenged a room full of could-have-been martyrs not to be half-Christians. Dad said that only living the faith was being a half-Christian. Being a full Christian meant passionately reaching out to those around them. If Ivory Coast was going to be reached for Christ, it was going to be through them. They were God's plan. Dad's message was simple. Being sold out to Christ with a small vision was being half a Christian while being sold out and driven by a large vision was being a full Christian.

I know many top theologians would argue this perspective. Being sold out is being sold out which is true, but Christianity is measured by more than just being sold out. Maturity is a key element of faith. An immature sold-out person will have limited skill and vision. A mature Christian has nurtured skills and honed the vision to be fully useful for Christ's kingdom. A person who has chosen to keep the vision small has decided to neither develop the skills to reach the currently unreachable nor nurture the flames for those around. A choice not to have a big vision is a choice to have a small vision. All into a degree is not all in.

Dad loved to take ideas to a full conclusion. Everyone had taken or was in the process of taking the Gospel to their family and friends. The calling of giving Salvation to their families and friends was urgent. That vision for their families and friends was not sustainable. After everyone in the calling was told then what? Without a new vision, the urgency of the Gospel was gone, and the faithful would sit on past victories. There would be no one to tell among friends and families. If the urgency of the Gospel died, so would the revival. Without an urgency of the calling, the warnings of being a half-Christian would have become true.

I did not know what kind of response there would be during the invitation. Once dad gave the invitation, people started coming. No decision cards were filled out that day. I don't know if there were any salvation. The people had decided that they were all in. There was no holding back. Everyone got up at the same time and moved forward. Most did not get very far. Once they could go no further, they dropped to one knee and bowed their head over the other knee. There was one big huddle at the front of the church. Four people remained seated. They did not move. The four Darnells sat and watched the magnificent touch of God. I felt awkward. It did not seem right that I did not go forward, but that day, God chose for me to sit and watch.

4 Rejoice in the Lord always! Again,
I will say, "Rejoice!"

-Philippians 4:4 (WEB)

I Moved on with my Life

I know I am changing the topic like an old man telling a story. We will skip the rest of the African stories for now and set the stage for more of the Yopougongare story. You can wait a few more moments. I had to wait 30 years.

I went to seminary and studied divinity so that I could follow in dad's footsteps. Part of the seminary is studying the largest churches. I learned that the two largest churches in the world were in South Korea. God has moved into South Korea with great power.

Second Largest Church

One day in about 2006, dad called me up and asked, "Do you know where the world's second-largest church is?" I answered, "Yes! South Korea." Dad did not correct me, but asked again, "Do you know where the world's second-largest church is?" I answered the same way again, "Yes! South Korea!" Dad asked the same question a third time. I figured that the same answer again was not going to get me anywhere. This time, my childhood came flooding back to me. I answered, "Yopougongare!" Dad answered, "Yes!"

Dad explained that the church was running about 120,000 thousand people. That is a lot of people. Imagine lining that many people up where every person could be given a two-foot span to stand in. This is not two feet between each person, but a two-foot span including the body. A line of two feet intervals would be 50 miles long and loop back directly in front of us. 120,000 with a two-foot span to stand in is about 100 miles in length. That is a long line.

The church has since built a large auditorium which I do not know if it is adequate or not. The church has sent out over 250 missionaries throughout the world. The church is still under the cell concept where some of the cells have been cast throughout the world. It does not matter how many sit together at one time. What matters is reaching the world for Christ. They considered themselves one church with Dion Robert as their head pastor.

Those numbers presented are significantly out of date. The church is not much into the web. They do not call themselves Yopougongare. The family named it that from the beginning. I heard at one point that the church has over 300,000. The numbers don't matter. If this ever gets published, I may look for the most recent numbers.

28 We proclaim him, admonishing every man and teaching every man in all wisdom, that we may present every man perfect in Christ Jesus;

-Colossians 1:28 (WEB)

The Families Received Christ

Dad slowly lost the ability to express himself as he grew old. One of the last things dad told me was about the African families. I was still devastated that so many had lost their families out of persecution. One day, I asked dad about what happened to the families and the Christians that had been expelled. Dad just laughed. I realized then that I did not know the rest of the story. Dad said, "When someone would get kicked out of the family, the churches would put together teams to visit the family. The families often received Christ." One of dad's favorite expressions came to mind, "Oh my, my, my!" What a glorious thing happened. God highly rewarded these brave souls. Can you imagine being the one who brought salvation to the family and the village? Because of their bravery, Salvation has come! God turned their deepest sorrows into their greatest joys!

> 42 He commanded us to preach to the people and to testify that this is he who is appointed by God as the Judge of the living and the dead. 43 All the prophets testify about him, that through his name everyone who believes in him will receive remission of sins."
>
> -Acts 10:42, 43 (WEB)

JIM H DARNELL JR

Chapter 23

Baptism

Baptizing & Rain

Baptizing is a difficult challenge in some parts of Ivory Coast. The two rainy seasons are back-to-back for 3 months at most. For the other 9+ months, the country is dry. Many places have no places to baptize when the rainwater dries up.

36 As they went on the way, they came to some water, and the eunuch said, "Behold, here is water. What is keeping me from being baptized?"

-Acts 8:36 (WEB)

Baptism Meaning

Believers throughout the centuries have argued over the meaning of baptism. The arguments have centered on whether baptism saves, age, and meaning. Those arguments take on a whole new meaning in the lands of persecution.

Being curious about Christianity is often not seen as a big threat. Going to church is a lot less likely to get people kicked out of families. Christianity is often seen as a plus when it is mixed in with former beliefs. Sometimes, people just try to cover all the bases.

In lands of persecution, Christianity is especially seen as a threat on two occasions. One has already been mentioned –

when God is the only choice. Rejecting the old to latch on to the Eternal is often seen as a cause for war.

The other threat is baptism. Baptism is often considered a declaration of war. These two phrases are seen as synonymous, "When you get baptized" and "When you declare war on the family." The persecution often increases from attending church to believing. However, many times, families do not reach the rupture point until baptism. Baptism is often seen as saying, "I am one of them and now I am rejecting my family." Baptism is seen as an all-out declaration of war.

The persecuted believers esteem baptism so much. Baptism has a different meaning to the persecuted believer than it does to the family. Baptism is shouting out, "I will love my God above all else!" Baptism is also a confession, "I will suffer any price for my Lord!" The families get these first two meanings, but there is a third.

Baptism has a meaning that non-believers do not understand at first. Baptism says, "I have chosen to love my family so much deeper than ever before." Believers place their own lives in grave danger to tell their friends and families about Christ.

Baptism is a declaration of war. The family believes that the war is declared on them, while the believer is declaring war on the spiritual bondage that holds the family. What an irony! Christian persecution happens when the perishing is declaring war on the one who is bringing salvation. While often going to church and becoming a Christian is tolerable, baptism is seen as a declaration of war.

God calls, "Are you going to live for me or not?" Those who say a salvation prayer and refused to be baptized are saying to God, "I want You to die for me so that I may live forever, but I will not take up a cross for you. God, I need you to suffer for me, but I will not suffer for You!" Being baptized in a hostile land is a declaration of allegiance to God.

*16 He who believes and is baptized will be saved;
but he who disbelieves will be condemned.*

—Mark 16:16 (WEB)

A Snake in the Village

Every day, African women sweep their compound. They have homemade brooms that do not have a handle. The dirt is brick hard, and they sweep it vigorously. No sand or dust should ever remain on the compound. The sweeping also keeps any vegetation from growing as well. The compound was a lot bigger than the village itself. One day, it was obvious why the grounds needed to be nothing but hard dirt.

Most of the snakes fell into one of two categories. The two main categories are neither that it is poisonous or not. The two main categories are the time to say last prayers, and no time to say last prayers. The question when bit by a snake was not if the bite is deadly, but how long before death.

One day, we were driving by an African village. A huge snake had wandered on the compound. It was surrounded by many African men. I don't know the kind of snake that it was, but it does not matter I suppose. Their expressions made it clear. The snake was slow and about 15' long or a little longer. It was more than twice the length of the men. The men had some branches and a few things to throw at the snake. The snake was too slow to move much out of the way. It seemed to be an easy kill. The Africans could neither hit the snake with their branches nor hit it with the rocks and smaller sticks.

I saw terrified men too gripped with fear to be able to aim. At the time, I did not understand. The Africans knew what kind of snake it was. They could tell countless snake stories. They could recount about when maybe it was their brother who died. Maybe another story could be recounted about their aunt's agonizing death. They could all tell stories of when they could have easily died.

These slow-moving snakes blend in so well with the environment that they are nearly invisible. They are so slow that they cannot chase down bitten prey. The venom only allows the prey to take a few steps before convulsions take over.

Dad did not stop the car. I watched the scene until the drama disappeared from my view. Everyone was too far away to get bit. They were so close that they were terrified, yet too far away to kill the snake. What a horror it is to live with every moment of fear of getting bit.

14 As Moses lifted up the serpent in the wilderness, even so must the Son of Man be lifted up,

- John 3:14 (WEB)

Baptism Minus One

Many times, even when a local church had a pastor, they all wanted to be baptized by dad. They wanted to be baptized by the one who brought salvation to them all. Sometimes they had to wait a while for baptism to occur. This particular group had to wait a while because we had been on furlough for a year in the United States. They often had to wait until the end of the rainy season so that there would be as much water as possible.

One year had been dryer than normal. Dad was able to call the pastor and talk to him about the baptism. Dad was concerned that there was no place to be baptized. During the phone conversation, dad asked, "Do you want to postpone baptism until there is more water?" The church pastor said something similar (translated expressions), "Not on your life, Pastor! We have waited a long, long time. We are not waiting any longer." Dad questioned, "Are you sure there is water?" The voice on the other end said, "I am sure there is water!" The group was so determined to show God their love through baptism.

Upon arriving, the group was so excited for their day had come. They had risked it all. They had suffered scorn and rejection. Their eyes were off their troubles and on their Savior. As dad was driving up, he realized that the area had not received much rain that year. Dad asked, "Are you sure there is a place to baptize?" Many voices reassured dad, "There is plenty of water to baptize!"

They loaded up in the car to find a place. The leader was giving dad all sorts of directions. When they arrived, dad was still looking for the water. When we think about water, we think of a lake or river. Because we know that their bodies of water were seasonal, we would think of something more like a stream or a pond. This place neither had large nor small bodies of water; they possessed a small mud puddle. Dad looked at that water, and asked, "Do you have anything else?" They answered dad with great confidence, "We have more water down the road a little bit."

They drove a long way and stopped. The people were right; there was water. Dad realized that a mud puddle was all that he was going to get. Perhaps this second one was better than the first one. The believers reassured, "See! We have water! There is no reason why we should not baptize."

They all got out. Dad had on an African suit. He got out in the deepest part of the mud puddle. The bottom of this puddle was very muddy. It was deep enough to immerse. Dad realized this was their dream, so he called the first lady up. She stepped out into the mud in front of dad. In immersion baptism, the one baptizing needs to take a step to lay the baptized one all the way down into the water. While dad was saying a few words before baptizing, both he and the woman were sinking further and further into the mud. When it came time to take a step, the feet did not move. Dad's feet were stuck in the mud. Stepping was no longer possible. They both fell over. Dad lifted his head out of the water and looked at the woman. There was still a part of her body that had not gone under. Dad reached out and poked her under. Men came into the water and lifted them both out.

I was very curious about the second one baptized. I asked Dad, "Did you ever figure out how to take a step or lay the person down without taking a step?" The answer was simple and clear, "No!" The mud was just too unstable to keep balance, and the mud made it impossible to take a step.

Everyone was baptized in like manner. Dad would position himself in the deepest spot. Someone would join him in front. Dad would say a few words. Both would fall over. Dad would look to see if they went all the way under. The parts still out of the water, dad would push under, then some of the men would come and pull them out.

These are beautiful memories. The people were so eager to get baptized that the hardships were nothing but an adventure. Kids sometimes like to get muddy. Adults rarely play in the mud. Please forgive me if I am getting too sexist. How many women in your family would be beaming with joy when they are covered with mud? It was far more than baptism that brought the overwhelming joy. The Spirit of the living God had come! The Eternal One had reaffirmed to them all, "You have done what I wanted. You have been faithful to me, and I am proud of you! Today we celebrate your love for me." God's wondrous touch made this a highlight of their lives. This was one of dad's fondest memories in Africa.

When the baptism was over, everyone had one thing to tell dad. You know what it is, and you could say it with me. They said, "See, we told you there was enough water to baptize!"

22 But Jesus answered, "You don't know what you are asking. Are you able to drink the cup that I am about to drink, and be baptized with the baptism that I am baptized with?" They said to him, "We are able."

-Matthew 20:22 (WEB)

Baptism and a Snake

Some places had more water than others. The issue at this place was not the water. Dad had gone out quite a way into the water until he found a suitable place to baptize. Dad had decided on an ideal place. Dad turned around and called for the first person. A lady came out. She waded out until she came right in front of dad.

The people on the bank were extremely excited. They were holding on to trees and leaning over the water. They were so excited that they were captured by the moment. Dad noticed that a snake was on the bank's edge. Many people were leaning over the snake. Dad grew concerned, and he tried telling them that there was a snake. The lady saw the snake as well, and she went to making all sorts of fuss. Dad was trying to talk over the lady while she was squealing in terror. Communication was lost in the moment.

Dad kept pointing. Suddenly, the people realized that ruckus was not a baptism celebration. The excited people on the water's edge looked down. Their celebration ended, and fear gripped the bank.

Everyone started picking up things to throw at the snake. The snake turned around and stared at the people while the snake swam backward. If you thought that the lady by dad was terrified before, she found a whole new dimension of fear. She was screaming at the people on the water's edge. Dad was yelling at them, "Stop! Stop throwing things!" The two in the middle watched the snake back up in their direction. Both the lady and dad were unable to get the throwers to pay attention.

Dad and the lady had nowhere to go. They did not know exactly which way the snake was going to go. All they knew was that the snake was backing up exactly in their direction. The people on the edge never hit the snake, woman, or dad. The snake did not realize that it was backing directly into two people. The snake backed up and hit dad in the arm. Startled, the snake turned around and swam away. The snake disappeared from their view. Death was still out there, but where they did not know.

The focus on baptism had been lost for the fright of death had visited. Dad had a choice. What was he supposed to do? Calling off the baptism did not seem like the thing to do. The woman was highly wound up and needed some time. After a little bit, she calmed down. Dad went on to baptize her and the rest of the group. They all got in the water and took their turns.

Faith in God does not diminish the fears and pains of life. What faith does is give love, joy, and peace that is greater than the troubles. Faith does not eliminate troubles. God sweetly and gently carries those who walk in faith.

On that day by the water's edge, everyone but dad may have been a new believer. Even the veterans had not been believers that long. God is in the business of transforming fears into faith. God asked them questions like these, "What would have happened if you had died?" The Africans could only have responded with something like, "We would have begun eternity with You." God has more questions in times like these, "Why did you live? Why did you not slip and fall in the water as others tried to get away from the snake?" All they could answer would be, "Because You chose to give us another day!" God has a way of asking more and more specific questions, "Next time, if you are in harm's danger like that lady the snake was backing into, will you trust Me with your life during the event?" God asks some tough questions that take months and years to process. For those who grow in faith, the answer is always the same. God renews strength. God transforms the mind and floods the heart with love, joy, and peace. Once God transforms a soul, the emotions still run very high during troubles, but the Glory of God reigns. While the troubles of life attack, the beauty of God shines through the trusting believer. God is very excited about His plan.

33 I have told you these things, that in me you may have peace. In the world you have trouble; but cheer up! I have overcome the world."

- John 16:33 (WEB)

Baptism and the Log

People normally don't think of a log and baptism as going hand in hand. In this story, the log is very much a part of baptism. Many times, dad has performed baptisms in less-than-ideal conditions. To the Africans who have risked it all, the only not ideal condition is not getting baptized.

Once again, dad was asked to go baptize with far from perfect water conditions. Dad asked, the same question to several different groups seeking baptism, "Are you sure you have enough water to baptize?" Dad was assured that they had enough water to baptize.

On one of these baptismal trips, dad was directed to a small shallow pool of water. It is called a pool of water because the water came mid-thigh. The water was like all other places so thick and dark nothing can be seen beneath the surface. Finding a spot to baptize within a puddle or a pool is not always that easy. The pool of water is just like any other flooded overgrown area. All sorts of grass and shrubbery reach beyond the water to find air. Areas often must be cleared out some before baptism.

This pool of water had a small clearing in the middle. The clearing was occupied by a log. Dad tried to find space on one side of the log. Dad bumped his knee into the log. Dad thought in his mind, "This won't work!" So, he went to the other side of the log. Dad bumped his knee again. Dad was surprised that he bumped his knee into the log again because that is not where he thought the log should have been. By now the log had enough of dad, and the log just sank. Dad did not know where the log went. As soon as the log went underwater, it became invisible because the water was so dark.

Dad did not know what to do. Should he baptize? Should he not? He was reassured many times that they had a wonderful place to baptize, so dad found his spot just where the log was. He called out his first person and began baptizing. One by one dad called them out. He said his words and laid them back into the water only to lift them again. Dad continued the sequence until all who wished were baptized. Throughout the story, dad kept calling the big thing in the middle of the water a log. Ivory Coast does not have sinking logs. Logs don't float for a while then all of a sudden get heavy and sink. We all knew what we thought it was. Ivory Coast does not have alligators like the United States; instead, they have crocodiles. Crocodiles do not usually grow quite as large as alligators, but they often have a meaner streak. At least, that was the reputation that crocodiles had. Crocodiles do find humans a lot tastier than alligators do.

As dad was telling the story for the first time to me, I was burning within with a question. I waited until dad had finished before asking, "Did you ever tell them that there was a moving log in the water?" Dad said, "Of course not! I was reassured many times that they had a great place to baptize, so I baptized them. They were right! They had a great place to baptize." Dad answered my question. Dad never did tell them that there was a crocodile.

We don't know how big the crocodile was, but we have some pretty good clues. The log was big enough to stop dad's progress and give his knee a fairly good thump. Had the log been smaller, dad probably would have pulled or pushed it out of the way instead of trying to walk around it. In my dad's account, the crocodile was more than big enough to be life-threatening.

Enough stories like these can be sobering. Both the snake and the crocodile were great predators in the water. The snake was so quick in the water that it could have bit dad if it wanted to. The crocodile was probably big enough to manhandle dad if it had so desired. Dad had taken away the best spot in the water from the crocodile. Both the snake and the crocodile were provoked enough to strike back.

Dad never seemed shaken for dad had made his choice. He was going to be about the Kingdom's business. Dad continued to baptize as the years passed without whining, complaining, or boasting. Dad was not reckless. Dad saw a need, and dad followed God's leadership.

Imagine mom for a moment. She had prayed for a Godly husband who would know the will of God. Her faith was put to the test. After the snake, would she object to her husband baptizing? Dangers became more evident. After the snake and the crocodile, mom's response was still the same. God was in control, and she was still trusting God to lead her husband. Mom knew that because God had spared her husband in the past that the future was still no guarantee. Any event could have seriously hurt dad and could have been his last, but mom had made her choice also.

13 For we write no other things to you than what you read or even acknowledge, and I hope you will acknowledge to the end,

-2 Corinthians 1:13   (WEB)

Chapter 24

Food

African Food

Dad ate a lot of things in Africa. Most of the time, dad ate graciously. Sometimes the food was not to dad's liking, but he ate anyway. The Africans were very generous in sharing what they had. Let's set the stage to go back to his childhood years.

41 For whoever will give you a cup of water to drink in my name, because you are Christ's, most certainly I tell you, he will in no way lose his reward.

-Mark 9:41 (WEB)

Dad Faced his Bullies

You would think that a brave man like dad would always be strong. Dad stood up to bullies that were most likely going to give him a good whooping. Dad had only one answer. If he got a beating, he was going to take it. The bullies were usually bigger, stronger, older, and in a group. Dad was alone, smaller, younger, and weaker.

Many times, the bullies were so surprised by dad standing up to the challenge that they would back off and walk away. Other times, they just beat dad. He tried to defend himself, but it was impossible.

Sometimes after a beating, dad knew that the bullies may start again. Dad still only had one choice. If dad was bullied again after a beating, he would still get up into their faces. Dad chose those beatings were more acceptable than being bullied. Dad had learned the hard way that bullying was worse than a few beatings.

When dad grew old, he grew more tenderhearted. He was always willing to stand up if he had to, but he had compassion for his adversaries. Dad chose to be fearless.

14 For as many as are led by the Spirit of God, these are children of God.

-Romans 8:14 (WEB)

Dad had a Weak Stomach

Dad was a strong man until there was a little spot of blood or throw-up. As soon as he saw those splotches, dad would start to gag. Dad would try awfully hard not to throw up. Many times, dad would turn away. If turning away was not enough, dad would have to go to the restroom.

The Rest of Us

When the rest of us saw something gross, we had two problems. We saw the thing that grossed dad out, and we had something far worse. We had to listen to dad. Everyone could not gag – what a terrible mess that would be.

TapeWorm

Dad never told this story much, this got to him more than anything else I remember. Dad never thought it was funny, but the rest of us laughed a lot. I guess this is one way I honor the dead. This story so grossed dad out. Here goes!

Dad was ever so embarrassed about the tapeworm incident. One day, dad was sitting on his throne, and he had an unusual experience. As he was releasing a bomb, it felt very different. Dad peered within the toilet bowl and saw a long tape-like thing. I don't know if dad thought his innards were rolling out or what. Such an occurrence would almost certainly mean death. It was like something he had never seen before.

The thing had to be pulled out of the toilet from among the poop to show the doctor. I know one thing for sure, dad did not scoop it out. A lot of discussions happened before flushing happened. The funny part was not that dad got a tapeworm. The funny thing is how dad would carry on. Go figure this out. A snake ran into dad, and dad was okay. A crocodile lurked right beside him in a small pool of water, but dad was okay. Dad would get beat up, but he could carry on. A tapeworm, vomit, or a drop of blood, and dad was hysterical.

Visiting an African in His Home -- Chairs

Africans knew no greater honor than to have us come into their homes. They were always as gracious as could be. Sometimes, they were so poor that they did not have but a few wooden upright chairs. Sometimes, they would only have like one chair. two, three, or four chairs were often common. As soon as we would arrive, they would ask their neighbor for a chair. I wondered if, at times, we sat in all the chairs of the immediate neighborhood. Sometimes the chair was so wobbly that sitting down and getting up had to be done gently.

Often, there would not be enough chairs. Everyone had to sit in one; that is, everyone who was a guest. I was an elementary-age boy. I got a chair, while the rest of the adults often stood. I had ridden in a car to get there; I had already sat plenty. This was a land where adults were more important than children; I sat while the adults stood. I did not get it, but I appreciated the love. Still, I politely sat while the hosts beamed.

14 For the whole law is fulfilled in one word, in this: "You shall love your neighbor as yourself."

-Galatians 5:14 (WEB)

Rock Candy Craze in the United States

When I was young, rock candy was the craze. Rock candy was the size of slightly overgrown grains of sand. These little nuggets were to be thrown into the mouth. Suddenly, almost nothing would start to crackle and pop. The sound in the mouth was quite impressive, but others did not hear much. It was like private mouth fireworks. The little rocks would transform into a mouth full of bubbles. Rock candy is quite an experience. Many of you are familiar with rock candy and know what I am talking about.

Visiting an African in His Home -- Sodas

It would have been a great shame if we went to their houses, and they did not serve us anything. Most did not have anything appropriate to serve. They knew that much of their food and water would make whites sick.

Their favorite choice to offer us was a soda. The sodas tasted like sodas, but they were very different under the circumstances. As I have said many times before and will say it again – that was the tropics. Most neither had air-conditioning nor a refrigerator. Saying the soda was not cold is not accurate. The soda was often sunbaked hot. Some people like hot tea or hot coffee; the Africans are served hot soda. Most of the Africans were too poor to drink sodas, so they did not know what they were serving. Hot soda had the same effect that rock candy had. All of the bubbles popping in the mouth were fascinating. Hot sodas just seem to explode in the mouth – cool.

Before I sound disrespectful, let me assure you that I was overwhelmed by their generosity. The Africans did not know what they were serving. They did not know that sodas should not be left out in the sun. At some places, the cokes were just room temperature hot which was still quite hot because the sun heated the building quite well. It was important to understand that their lack of familiarity with the sodas meant what they gave us was incredibly special.

It was not uncommon for the man that my dad was visiting to say that he lacked a portion to pay for the soda. Dad would have to pay for part of the soda. You see, they had to offer us something, and they did it with great joy.

Many would come by money very sparingly. In the village, they would grow their food. There was always so much to buy with so little money. The host would spend all he had on us just to give us his best. The generosity was truly humbling.

Once I talked to dad about these home visits. Dad said that their status in the village would often go up dramatically because they had such important guests in their homes. Having us in their homes was such an honor to them.

The Africans were just generous people, and it showed on many occasions. They would give us all they had to live on. They would let us sit in their own and their neighbors' best chairs.

23 And whatever you do, work heartily,
as for the Lord, and not for men,

-Colossians 3:23 (WEB)

Eating at Koumassi Church

One Sunday, the Koumassi Church invited us to a special event. The church prepared a wonderful meal for us. The food was a very typical African meal. They served a rice base covered with a meat sauce. It sounds really good except for one problem. The Africans love to throw in very hot peppers by the handfuls. They love making the food pepper hot.

On that day, I was overwhelmed by the heat in the food. We were all brave and ate food that torched our mouths. The Africans were generous for they brought in five sodas for the Darnells. I remember taking my first bite; the pepper was blazing hot. I thought that a drink of my soda in a bottle would help. I was wrong. The bubble explosion made the hot, hotter. I decided then to save my soda until the very end. I dug underneath to find some rice that was not stained by the meat sauce and placed it on top. In this manner, when the rice and meat sauce combination entered my mouth, the meat sauce would be somewhat shielded by rice on both sides. Once I placed the meat sauce in my mouth, I swallowed. I just kept swallowing without chewing.

I just have to remember the day. The church tried so hard to please us. They gave us the finest they could give. They truly showed their love. They gave me a memory that I have laughed about all these years.

8 *Let them praise Yahweh for his loving kindness, for his wonderful deeds to the children of men!*

-Psalms 107:8 (WEB)

Ed and Dad in the Village

One time, Ed (another missionary) and dad had gone together to eat in a village. The village honored the 2 by letting them eat first in the hut by themselves. On that day, neither man could stomach the food. They just could not eat it. The dining area was a hut with a dirt floor. That day, they just dug a hole and buried it. The missionaries did not think that was their finest hour.

Potted Meat

I cannot make too much fun of Ed and dad for wimping out. I did it once. I was probably about 8 to 10 years old. I went on a journey with dad to an African village. They were serving food that was terrible because it was too flaming hot. I knew that the food would make me sick until the peppers came out of my system. The food was so hot that all the internal plumbing would burn in the process including the exit hole. I sat there a moment looking at my food. The Africans knew that I did not want to eat it. They bought me a baguette (French loaf as we call it) and potted meat. I was so happy. I spread a little potted meat on the bread and ate it with such joy.

Bones

Several times I have eaten with Africans, and I was not able to identify the bones. The Africans ate everything. They ate mice, rats, snakes, and birds. Snakes and birds sounded good, while mice and rats did not sound quite as good. They cooked their meals with whatever meat they could find. Sometimes, all they could do was to get enough meat to flavor. They cooked with the bones. I remember pulling out bones that looked like little ribs. I had no idea what the critters were. Some little curved ribs could just as easily have been a rat or a bird.

I wish I had known the bird test back then. Bones can be snapped to look inside. If the bone is hollow, it is a bird's bone. If the bone were not hollow, I would have figured it was some rodent. Some of you may be grossed out. Others are laughing so hard to think that is even a big deal. If I had eaten bird, then that was normal food. If I had eaten rats and mice, then that would have been awesome. I would have had something to brag about.

Once we were driving down a dirt road. Directly ahead of us was a pick-up truck with a canvas cover. We were not aware that there were two men in the back of the pick-up. Suddenly, someone pulled back the canvas flap. A man lifted a rat and hug it over the edge to let us see his trophy. The rat was a little longer than the height of the tailgate and that was not counting the tail.

The Pig

One time dad was invited to a marriage. Having dad attend was such an incredible honor that he was given a little pig. Dad wanted to leave the pig behind because he did not know what to do with it. The family was very insistent that dad took it. They put the pig in the back of the station wagon. Now, pigs are good at two things. Pigs are great at squealing like they are dying. Pigs are also great at making a mess when there is not one.

Basic vehicles back then were quite bland. We are very thankful that the metal frame was not carpeted. The back seat had a plastic cover. All the pig could touch was plastic and metal.

The trip home was a fairly long one. The stress caused the pig to have lots of squirts. Much of the road was dirt and rough. The squirts slickened the metal floorboard. The pig would step into the mess and flip due to the rough road. The pig rolled many times in its droppings. The pig changed colors that day. The little creature leaned up against the plastic coverings and smear painted the back of the car. I know one thing. I am thankful that the mess was not mine to clean.

We fattened the pig up a little. We were then very happy that we kept the pig. These are memories of adventures together.

Foufou and Foutou

These were quite delicious. We went and ate with a family.

They decided to make traditional African foods out of plantain bananas. You have probably often seen images of African women using a pestle and mortar (a huge stick that is slammed into a wooden bowl). The plantain bananas are pulverized into mush. The taste was good. The food did not chew. Normally during chewing, food breaks apart and much of it just disappears down the throat. This food did not disappear. After chewing for a while, the glob was still intact. During chewing, the mouth emits a lot of saliva to coat the food. Chewing had only one effect on the glob, for the glob swallowed the spit and grew bigger. Dad was amazed for he said, "You can't chew it; you just have to swallow it!" These foods were quite delicious.

Roadside Plantains

This is some of the finest eating on the whole earth. It is so unhealthy. The plantains are fried and are sold out of the huge cone skillet. The whole family was eager to eat plantains.

Roadside Coconuts

When coconut husk turns black or a rotten brown, the meat (the white nut) is ripe. However, on the roadside, people stop for a drink of coconut juice. The lighter the color of the outside coconut husk, the less meat there is. Instead, the inner shell is filled with coconut juice. Not only is there a lot more juice when there is no or little meat, but the juice also tastes much better.

We would stop by the side of the road and order a coconut. The lady would use a machete to slice off the top of the coconut. Drinking the juice was messy. We would spread our feet apart and place the coconut up to the mouth. We would turn it over and drink. Juice would seep out and roll down the chin. The juice flowed safely to the ground without messing up the clothes. On a blistering hot day, the coconut juice was very satisfying.

Sweetened Condensed Milk

I was surprised that Africans would sell cans of highly sugared milk by the side of the road, but I soon found out why. The immense heat would make people lightheaded. The dizziness was caused by the draining of the liquids and the zapping of calories. A small can of sweetened condensed milk took the dizziness out of the head.

African Fruit

Africa has some of the best fruit in the world. To this day, I miss the fruit. Mom would sometimes buy me vine ripe tomatoes by the 5-gallon bucket. I would snack on tomatoes all day long. Mom preferred to serve things like fresh green beans with the meals.

Fruit Bats

In Abidjan, we had fruit bats by the millions. The bat's bodies are about the size of a dove while the wingspan is six feet. These bats loved fruit trees. They would completely cover the bottom side of the branches of the trees. They would squeeze together as tightly as they could fit. These big branches would just sag under the weight. Once, a branch about two feet thick snapped under the weight of the bats.

For a while, my friends and I would hunt the fruit bats with our BB guns. My friend, Rob Simrell stored the bats in his mom's freezer. After a while, when we had several, Rob cooked them out on the grill. I had never eaten one, and I could not be a chicken, so I had to join in the feast.

The fruit bats were surprisingly delicious. These bats ate fruit all day long. The sugar in their diet made the fruit bats taste like sweet chicken. You ought to try some someday.

The World's largest Open Market

Abidjan had the world's largest open market. It was huge. I

went with mom willingly, but it was a little overwhelming. One of mom's ministries was to demonstrate love to the Africans, but this was over the top. The open market was so big that I would lose my way in, but that was okay because mom stayed close enough to guide me.

The only part I disliked was the first minute inside the market. Mom would always give me a bag for the produce. There were always unemployed men standing by the entrance. They had no way to make a living, so they hung around for every opportunity. As you can guess, they saw me as an opportunity. They would grab the bag in my hand and try to yank it out. Mom would walk on and ignore the terrifying moments. I was terrified of these men. They would shake me like I was a rag doll while yelling, "Let me help you! Let me help you!" I did not like it very much. I realized that I was so totally helpless to protect myself. I did not know if they were going to "help" or steal. Help always happened for pay. We would never get the bag back until we paid them for their services. If they ever did yank the bag out of my hand, I would have felt great shame because I would not have been big enough to help mom during those elementary years.

After it was obvious that I was not going to let go, some of the merchants would yell. Leave them alone. As quickly as it began, it was over. We were free to walk. Mom would weave her way to her customary merchants with whom she had already agreed upon the prices.

Mom did not just want to go to the edge of the market, mom wanted to go to the middle. At any one time, the market would be 10,000 people. We are talking to lots and lots of people. I always noticed one thing. No matter how many blacks I saw, we were always the only whites. Mom's point was that she was willing to break down social barriers and love people.

The ground was dirt and very uneven. During the dry months, the holes were fascinating. During the wet months, the holes could not be seen. Much of the paths were underwater. Not all the ground underwater was the same. Some areas were very shallow, other spots were quite deep. I was thrilled that I would wear my water boots to the market. That was a lot of fun. Some of the holes were so deep that they could have flooded my boots. I was very thankful for the dry visits when I learned where the deep holes were. I had played in water many times. From stagnant water coming over my boots, I often got ringworms and other such rashes. The rashes were always curable, so it did not matter much. Mom was right; we were quite safe in the market because the Africans looked out for us. There were a few obnoxious people while most were super kind. I wonder what I would have done in those people's places with no way to make a living. They were eager to make a life without any opportunity.

22 To the weak I became as weak, that I might gain the weak. I have become all things to all men, that I may, by all means, save some.

-1 Corinthians 9:22 (WEB)

The Wildlife Game Preserve

Ivory Coast had a wildlife game preserve. Mom and dad thought it would be really neat to go and visit it. They took us to see the wildlife, but wildlife was impossible to see. The wildlife was under the canopy. The canopy was a huge box. The tall trees fought for sunlight way up high. On each side of the road was a vertical cliff of green. Nothing could be seen beyond the green wall.

There were animals. Once we heard of a herd of 100 elephants trampling a banana plantation. Another elephant had grown so old that its tusks dragged the ground. It died a few days later of old age. The animals were there, they just could not be seen.

After driving for a while, I grew tired of looking at nothing. I asked dad to stop and let me run by the car. Dad agreed and let me out. Dad had other things in mind. As soon as I closed the car door, dad took off. I knew that he thought it was funny.

I was so eager to run until dad took off then I could hardly find the strength to run. I was surrounded by flies as big as bees. Every time I stopped running, the tsetse flies would bite. I wanted to run. I did not want to run. I was concerned about where I was. We were in a game preserve. There were supposed to be lots of big animals. The kind of animals that would eat little boys.

About one and a half miles later, there was the car tucked under a hilltop. I asked Dad, "Do you know where we are? We are in a game preserve." Dad said, "Oh, I forgot about that." This story has fallen in the food category because I was concerned that I might become food for some animal.

Cattle Drives

The tsetse fly carries sleeping sickness which is extremely dangerous for cattle. For that reason, the lands nearer the coast were not used to raise cattle. A lot of African ranchers would drive their cattle toward Abidjan. These ranchers would drive their cattle very slowly. We are not usually talking about big herds, just lots of herds of all sizes. Cattle would often travel 200 to 500 miles just to reach the coastal market of Abidjan.

The cattle would use the same roads as cars did. The cattle felt like they owned the road. Slowly cars would edge their way through the cattle and continue.

Meat in the Open Market

The beef that reached the open market had several less

favorable characteristics. The cattle were sold directly from the rancher to the open market. The cattle did not spend time in feed lots. Feedlots are places where cattle do not move much, and they put on a desirable amount of fat. Out-of-shape cows have more tender meat. Cows from cattle drives have tough meat with almost no tasty fat.

Let's visit an old theme again. This is the tropics without air-conditioning. Butchers in the open market hang cows by a rear hoof. The meat just hangs there, and the temperature outside is just sizzling. Mom never bought bad meat. I was amazed.

The purchasing process was simple. The butcher would ask mom what kind of meat she wanted. He would point to a place on the cow and mom would agree to buy meat from that spot. He would angle his knife to show thickness. Once again, mom would usually agree because he knew what she wanted.

He would slice the meat just like she wanted. Let's back up again; this was the tropics. Buying meat was not quite that simple. The cows hanging in the open market were never seen. Mom went to open-air butchers, and this was the tropics. Flies multiply like crazy over there. As I said, this was the tropics where the cows could not be seen hanging and flies multiplied like crazy. You know what you think I am saying. Yes, the cow hung by the hoof could not be seen because it was so covered in flies.

For the butcher to display the meat, he had to shoo the flies. One shoo was not enough. If he wished to display the meat, he had to keep waving his arm.

Mom had the marinade sauce that she concocted just for the meat. The acids from the fruit would soften the meat. Mom would leave the meat in the refrigerator tenderizing it in the juices for three or four days. Afterward, the meat was ready to eat.Mom never focused on the flies or the toughness of the meat. Mom chose to go where other whites would not. Mom could have been grossed out by the flies or been fearful for her life going where other whites would not. Mom could have done all her shopping in the supermarkets and bought higher-quality meat. She just chose to live for God and let people know that she loved them.

Flies

Flies were everywhere. Every day the house would fill with flies. When we ate, I did not mind the flies on the food very much. No one ever seemed to have become sick from flies licking the food. I did mind when flies were annoying and would crawl all over me. Fly control was a major chore. Many times, when the family was getting ready for the meal, my job was fly patrol. Before every meal, there would be 50 to 100 more flies. It takes a long time to kill that many flies. Besides, fly-swatting was a lot more fun than assisting mom in the kitchen.

JIM H DARNELL JR

Chapter 25

Family

Mom

I have three repeated memories of mom for she did these things all the time. Mom would work all the time. She worked hard and worked a lot. There were always more chores to be done than she could finish. A second memory is mom would sit in the living room with her green French Bible open. Mom spent a lot of time reading her Bible. She chose to read it in French because she needed to be able to express her faith and do church work in French. Mom spent a lot of time with that Bible. The last memory is she getting on her knees and praying. She had trouble with that one. She had surgery on one of her knees when she was young, and that knee always stiffened up. Mom worked for her God and spent time with Him both in her French Bible and on her knees.

9 Even as the Father has loved me, I also have loved you. Remain in my love.

- John 15:9 (WEB)

Patty

Life was quite different for Patty than for me. In Ivory Coast, the women were expected to work after school when my African friends came over and played. Patty did not have the opportunity to make female friends. Patty did not gripe, but I could tell that she wished for friends as I had.

In many ways, I knew Patty best. We were just a year apart and we spent a lot of time together. When I just turned 14 and Patty, 15, we went on a journey together. We left home to go to boarding school. The three years created a strong bond between us. The school did not have a phone, so we had no contact with Julia, mom, and dad.

Patty looked out for me. I needed a lot of help. I had grown up without the ability to express myself. It is not that I did not want to express myself, I just couldn't. If I wanted to say anything at all, I would have to spend days rehearsing what to say just to spit out 4 or 5 sentences. Words just did not come for me for I thought mostly in pictures in those days. If I ever tried to say something out of the ordinary, I would start and just stop. Patty would say just enough to get me going again until a few words later when I was stuck again. The whole family tried to help me without taking over, and I am very appreciative of them all. However, it was Patty who was there more than the others and kept me feeling safe.

One day, Patty's class from boarding school went down to Abidjan to swim in the ocean. Her best friend was swept out to sea. She stayed out there a while. She could not figure out how to get back in. The school would not permit anyone to attempt to rescue her because they did not wish to lose more students in a rescue event. Eventually, her friend grew tired and slipped beneath the surface. Patty never did say a whole lot about her, but she was much sadder for a long time. I did not know what to do or feel. I was very confused. I did not know what I should feel. I felt like a failure because others expressed much grief, and I said nothing. Being silent upon her death made me feel like I did not care. Later, Patty and I chose the same university. I was glad to have Patty in my life. Especially in my younger years, Patty was there.

8 He has shown you, O man, what is good. What does Yahweh require of you, but to act justly, to love mercy, and to walk humbly with your God?

-Micah 6:8 (WEB)

Julia

Julia is 6 years younger than I am. When I left home to go to boarding school, Julia was only eight. Up till that point, Julia and I often played together. I just remember having the fondest of memories of time spent together. We just played and had a good time. One of the deepest prices I felt as a missionary kid is not being able to spend much time with Julia when we were young.

My most striking memory of Julia happened at the yearly parade in Ivory Coast. People from all over would come to see the spectacle. The crowds pressed in so tightly. Thieves often took advantage of such events. Before one parade started, dad was carrying Julia on his shoulders. A man slid his hand into dad's pocket and took dad's wallet. Dad felt it and turned. The man took off running. Dad took off running behind him. Dad caught up with the man. Many times, thieves worked in teams where the one who runs was the decoy without the wallet. Dad did not know if when he caught up with the man if he would still have the wallet. Dad did catch up with the man. He was yelling, "I did not steal your wallet! I did not steal your wallet!" All the while his hand was waving up in the air with the wallet in it. Several of the African men around told dad, "Don't you worry about this; we will take care of it!" Sometime after the parade dad was telling the story about the man stealing the wallet. Someone who had attended the parade said, "I saw a man fitting your description who was all bloodied up. He was so beaten up that he probably died shortly later.

About the time dad regained his wallet, dad realized that Julia had disappeared in the process. The crowd was so thick that getting separated by a few feet could make someone not findable. Patty, mom, and I had stayed together. Dad was easy to find because of all the commotion. However, we had no idea where Julia was. We retraced our steps back to where dad was at the time of the theft. We were all be wondering if Julia was going to be there. When we came close, a tall African man was standing there with a little white girl on his shoulders. He knew it would be really difficult to find her unless she was above the crowd. Mom and dad thanked the man profusely. Julia was as calm as ever. I am very thankful that we got Julia back. On that day, the wallet was the decoy and Julia was the prize.

10 Yahweh's name is a strong tower: the righteous run to Him and are safe.

-Proverbs 18:10 (WEB)

Chapter 26

Friends

Kojo

Kojo (Koadjo) was my best African friend. He was always loyal. He had a profound influence on my life. He was so talented that in many ways he was my mentor. The next set of stories is either about Kojo or setting up a Kojo story.

The Necessity to Write Kojo Stories

The stories also build on each other. Part of credibility comes with ability. One of the stories is still hard for me to believe even though my eyes have seen it. At the same time, telling how I fit in the stories makes me very feel sick because I feel like I am bragging.

Galimotos and Kojo

Kojo taught me how to make cars out of chain link fences. The cars I built ranged from eight inches to two feet in length. The cars were wireframes with rolling wheels. The front axle would turn and connected to the front axle was a steering wheel. The steering wheel was about five feet long and would angle up to about waist high. The car could be pushed around and guided with incredible accuracy. They were my favorite toys.

The chain-link fence had to be pounded into a straight wire. Kojo taught me how to make both sides of the car identical.

The wheels would roll smoothly, and the front axle could easily be controlled while walking was no small feat.

I would often add seats and light fixtures. Sometimes, I would splurge and buy flashlight bulbs so that the car would have real lights. An oversized steering wheel was attached to the front axle. The steering column rose out of the roof of the car up to waist high. I always added a gear shift, blinkers, and a horn.

Inner tube Tire

This was back in the day before tubeless tires. Cars were all the time having flats. I did not realize it at the time, but it was the blowouts that interested me. When a tire went flat, the inner tube stayed within the tire. A blowout was different. The tire would shred leaving cars to drift on three tires. A shredded tire would also free the inner tube. As we drove along, I neither tried to figure my way around, nor was I not looking for girls, but I was looking for inner tubes by the side of the road. Every time I saw one, I would plead with my dad to stop.

The galimotos were nothing but wire. They were neither glued, taped nor soldered together. They were wound together with inner tube strips. I would cut the inner tube into strips. If the strips were stretched tight enough the nots at the end would hold.

I often ran out of the inner tube. As always, before I had the chance to build another car, another inner tube was waiting for me beside the road. I was never in need, nor did I ever have a large collection. I always had just one inner tube that I was working on. I had come to believe that God would provide not more than I needed, nor would I go lacking.

When Patty Was Almost Killed

The galimotos were fun to build until I ran out of fence wire. It did not bother me much because I simply believed that God would provide more fencing. I just did not have any idea of the cost of my new fencing.

The French hired African help around the house, and the French liked their vacations. They would go on vacation trusting their help to keep the house safe and in order. One family had two cars. Their best car was taken on vacation, and the other sat at home. The African who managed the house was alone with the alcohol. He drank so much that it impaired his senses. Once drunk, he had to take the other car out for a spin; oh, he spun the car all right. The African had two problems: one was that he was impaired from the booze, and the other was that he did not know how to drive. The alcohol extremely impaired his judgment. Not only did he take the car out that he did not know how to drive, but he also raced down the residential streets at full throttle. At one point, he lost control. He hit the ditch right in front of our house which launched the car up in the air. He was headed straight for Patty's bedroom, and she was in her room at the time. The only thing there to slow him down from crushing her bedroom was a chain-link fence. From making many galimotos, I learned that chain link fence wire bends easily but is hard to cut. This car must have gone high in the air for the story to unfold, I mean the fence to unfold as it did. The fence stretched with the car. At the impact site, the fence was only about one-third of the normal height. The fence never let go of the car. The car flipped and the fence just gift-wrapped the vehicle. Not only was there fencing on the front edge of the car but the fence was wrapped around the sides as well. The fence caught on the mirrors to create both a horizontal and a vertical wrap-around.

I had never heard anything like it before. Some of my friends yelled, "There was a wreck!" We all raced to see what happened. The whole family was home as well. We all raced toward the sound of the loud metal howling as the fence stretched and rubbed against the bullet. When we arrived, the driver had already immerged from the car and was walking around unharmed.

Patty kept saying, "I was in the bedroom! I was almost killed!" Mom and dad kept repeating, "Patty was in the bedroom, she was almost killed!"

I do not want to talk about my feelings. I just could not imagine life without Patty, so I just could not believe anything other than God was testing us and protecting us. I had a childlike faith in God. I had not yet learned that God does allow the worst to happen to His most faithful. Once again, I never said anything; I hardly said anything, ever. Everyone was visibly upset, and I was so quiet that it made me feel like I did not care. The wreck happened four or five years before Birdie's drowning death. I wished that I could have shown that I was upset. Because I showed no negative emotion, no one thought to see how I felt.

I also saw the fence. Most of the long fence was still usable for galimotos and not for much else. I saw God's hand of provision. Behind a shed was stored the damaged fence. There was always fencing to make galimotos until the end. As my time in Africa drew to a close, I had time for one more galimoto; and you can guess it, there was exactly enough wire and time to make one last car.

The Pliers

Kojo did not have the same quality tools to make his galimotos. He often used a rock as a hammer to flatten the wire, and he used a straight-edge razor to cut the inner tube rather than scissors as I did. He also taught me that the wire could be bent back and forth many times to snap it. The curled wire edge could then be hammered back out straight. Kojo faced many hardships from poverty.

One day, I left a pair of plyers outside. Kojo found them. Those pliers would have been his most treasured possession if he had kept them. I would not have known. Kojo came up to me and handed me his discovery. It was hard for me to take my plyers from his hands for they would have meant so much more to him than me.

That day had a profound impact on me. I learned that Kojo could be trusted and that he was loyal. That day, I learned that Kojo was the real deal.

The Man Test

African culture was tough in some ways. Bravery and ability were demanded from each other. Due to the harshness of poverty, the brave and the skilled were highly esteemed. Life could be so difficult that young men were driven to be strong-minded and to excel in skills. Guys had to pass maturity tests to be accepted as a man. Those who failed were ridiculed greatly until they passed the test. Some of the following stories are grounded in the maturity tests. Kojo thought I had to pass the maturity tests as well.

The 1, 2, 3 Game

I did not like this game at all. Kojo and all his friends played it. To them going to school was such an incredible privilege. Education was seen as their only hope to escape extreme poverty. All Ivorians I knew during that time spoke at least one tribal language and French. Kojo always spoke very correct French. French was hard for me, and I struggled greatly in school. I made a lot of mistakes. I did not like the game.

The game's rules were very simple. Every grammatical mistake made while speaking was first to be made fun of. Everyone would point at the grammatical offender. Out of the mouths would be jeering and laughter. This was still the good part of the game. The next part was the bad part because I was usually the one who made mistakes. The guilty party was to rush to the wall and pound it three times while counting one, two, and three (un, deux, trois). That was the easy part. The hard part was figuring out the mistake. If I had known it was a mistake, I would not have said it in the first place. My friends would then be able to hit me as much as they wanted until I corrected my mistake. It is awfully hard to correct grammar while friends are hitting on all sides hard enough to sting.

Marbles

Kojo taught me how to play marbles. He always wanted to play keeps. It was a terrible game for he ended up keeping my marbles.

The Distance Game

I liked the distance game a lot more than the 1, 2, and 3 game. Kojo would randomly place a rock, stick, or draw a line in the dirt. He would ask this question, "How far away is the ...? I became fairly good at this game, but I only won about twice. Usually, the item was several yards away. Most of the time he could guess the distance within about an inch. Too many times Kojo laughed at me because my guess was too far off. On rare occasions, Kojo would miss his mark by two or three inches. I would laugh at him because it was a terrible guess for him even when his guess was better than mine.

At the time, I did not know how this little game was going to be so central. The story is still coming of Kojo's most incredible adventure. A story that needed to be measured.

Soccer

Kojo also taught me to play soccer which was the game we played most. In the United States, there are three sports seasons per year. In Africa, we had one season. All year long we played soccer. All my friends were great athletes and outstanding soccer players. Soccer had its maturity tests as well. The standard for being a real soccer player was being able to juggle the ball 100 times on various parts of the body.

The first test that I passed was juggling the ball 100 times on the right foot. Of course, when I thought I was ready, I had to have an audience to prove my skill. The audience always added so much drama that the skill became much harder. There were a few other tests to prove greater skill. I had to juggle the ball 100 times on my left foot. After that came the knees and the head. I thought I had passed all the tests until they brought a new game home. We had a huge yard. They would place the soccer goal on the far end of the yard. The sticks were not much further apart than the ball. The goal was to see who was the most consistent. The game just became harder. They would tilt the goal so that the ball could not enter the goal directly for it had to be spun to curve into it.

Every soccer move Kojo learned he would teach me. One of the last few times I saw Kojo he did two stunts that I have not seen before or since. Kojo would tap the ball between his heels. Quickly tapping the ball between the feet is a low-level skill, but this was different. Kojo was tapping the ball between his heels behind his body while running at full speed. The only way for him to see what he was doing was to lean forward. The ball was untouchable by anyone else because it was so far behind him.

The second stunt was the most impressive. Kojo would spin the ball around him in the air while running at full speed. The ball would keep circling without ever touching the ground. He had three touch points where he would tap the ball and barely alter his stride. As far as I know, these stunts may be beyond world-class to being one of a kind.

Kojo liked turning his opponents into doggies. Kojo was just quicker than everyone else. Sometimes, Kojo would neither pass nor dribble the ball. He would stay in one little area daring his opponent to take the ball away. He would get the ball as close as he could to his opponent, yet the ball might as well have been on another continent for the ball was untouchable. Sometimes, Kojo would roll the ball over the opponent's plant foot while the other foot was reaching in the wrong direction for the ball. The whole time Kojo would be laughing and taunting. Eventually, the embarrassment would be too great. The opponent would be swinging to hurt with the legs and the arms. Kojo would stay in the location still teasing with an untouchable ball. Even with the swinging to do great bodily harm, Kojo would still manage to dodge the flying fists, and still, he was able to roll the ball over the plant foot again. Eventually, while lashing out in rage, the opponent would stumble. The livid one would then be on all fours. He would be crawling as fast as he could toward Kojo and the ball; the opponent had become a doggie. At this point, everyone would step in and break it up.

When we picked for teams, Kojo was usually picked first. Sometimes, they would pick me. I just did not realize how much Kojo had driven me to excel.

Our last soccer game is still so vivid in my mind to this day. For the first time, our friends wanted Kojo and me to be on the same team. There were seven of them and two of us. It was not very fair, but it is what they wanted. These seven would have been on track to have starred on nearly every high school soccer team in the United States for they were very good. Because there were only two of us, they let us start with the ball. I passed the ball to Kojo, but he did not pass the ball back. He decided to play his little doggie game with all seven. Kojo stayed in one little area. He just danced around with the ball. All seven together could not get the ball from him. I wanted to play soccer, but it was not possible. The 7-on-2 game turned into a 7-on-1. I was left out. I had enough and the game turned into an 8-on-1. I had long since learned that I could not steal the ball from Kojo, but I could keep him from going around me. With the help of the other seven, I took the ball away. I raced down the field and scored. The outcome was very predictable on that day. I was the only one who scored. When they had the ball either Kojo or I would steal the ball away before they could score. If Kojo stole the ball, he would stay in the same area and play his doggie game until I came to steal the ball away. If I stole the ball, I would take the ball through the team and I would score. Sometimes, I dribble around in a winding enough of a trail where I had to dribble through the same players more than once.

Pro Offers

When Kojo was 11 and a half, he told me about some offers. Kojo was so good that he would have been a superstar at any level. He was too young and little to play pro, but he already had the speed and skill to do anything he wanted. Some of the most prominent professional teams had offered to start paying him at an early age to lock up his skills. Kojo turned them down. He said that one time someone would get him and hurt him. He was probably right. He was so good that he would humiliate the best as he did in the doggie game. Kojo never pursued soccer as a career.

Our Last Time Together

I did not realize that this 7-on-2 soccer game would be the last time I ever saw Kojo. We went on furlough. A year later, we returned. Dad had built the first of three houses on the other side of town. We moved there. I did not know how to find him. He always came to me at my old house. He walked nearly every day two miles each way to see me.

Rob Simrell

Rob was one of the three best friends I had during childhood. His parents were missionaries. We spent countless hours together playing. Rob was super loyal and gentle. We loved to climb trees together and hunt bats. It was Rob who grilled the bats for us to eat together. Fruit bats were super delicious. One day, his father asked us to chop down a major branch on a massive tree. Rob and I were high and falling could have been extremely dangerous. Rob and I took turns chopping that massive branch with a machete. The branch was like a major tree sticking off to the side. We were concerned that the branch would snap up throwing us off.

Patty and I chose to go to Samford University. One day, I was walking across campus, and I saw a ghost from time past. Rob Simrell had enrolled in the same school. Rob had continued to play and be friends with Kojo. Kojo had given Rob a parting gift and a tape for me. I guess Kojo thought the United States was a small place and we would happen to go to school together.

The Python

On the tape, Kojo was upset with me. I was his best friend. One day I left and never contacted him again. I do regret that now. I have had other great friends, but how many friends can you find that make you feel like they live for you?

The tape told a hunting story back in the village. This story is so unbelievable that I have a hard time believing it even from Kojo. My friend who showed incredible skill and poise needed all his abilities to survive this story. For the story to be so believable, the teller had to be incredibly honest and accurate.

Kojo had returned to the village to be with his mom. One day, the men of the village were out, and some men from another village had come to see them. The women wanted to fix them a meal, so they sent Kojo out to hunt. Going with Kojo was a younger person whom Kojo called "Boy!" They went into the pristine old jungle that has stood for centuries.

Kojo was easing around a bush to see what was on the other side. Kojo was not the only one searching for a meal. A massive python struck Kojo in the arm to hold him still. The snake immediately started coiling around my friend. (In all my years with Kojo, I never saw him lose his composure or panic even for a moment. Kojo sought with all his might to become a real tough African man.) Kojo continued to tell me that he panicked for just a second and dropped his gun. By the time the snake had coiled around him twice, it was up to his thighs. The boy who went with him was terror-stricken and ran away. He left Kojo to die in the jungle. Kojo was madder at the boy than at me. Kojo grabbed the snake by the neck that is if a snake has a neck. Kojo yanked with his right arm ripping his left arm free. Remember that snakes have long needlelike teeth turned backward. A snake this size would have dozens of inch-long teeth all locked backward. The yank was far more than breaking free for the teeth did not give way; it was Kojo's flesh that gave way. With his left wounded arm, he reached down and picked up the gun. He put the barrel at the snake's head and blasted.

You might think that the story ended there, but Kojo finished with style. Kojo picked up the machete that the boy had dropped and cut the snake in half. He said that the snake was too heavy to carry. I was wondering how a snake could be too heavy to carry. But, if Kojo was small enough to be a meal then the snake had to be big. Kojo had to honor his mom by bringing the game home as directed. He was also loyal to the men who came to see them. Kojo said that the snake was about 11 yards long. At that length, the snake could have swallowed many Kojos and a boy or two.

My friend Rob said, "Oh, by the way, Kojo gave me the snakeskin. Do you want to see it sometime? It runs the length of my bedroom above the bed." Later did not do. I had to see the snake immediately. The snakeskin was nearly a yard wide at the center. It started tapering off. After the snakeskin started tapering off, it was cut. If the wide spot was truly the center, then the snake was about as long as Kojo had said. The head end was jagged. My guess is that a shotgun was used to kill the snake.

Vincent

Vincent was Kojo's younger brother. Both were incredibly loyal to me. I liked Vincent. Vincent had the positive traits that his brother had. Not only did I have a best friend, but a best friend's brother as a best friend.

Albert

Albert came with his group of friends. God had blessed me not with just one best friend, but a lot of friends who supported and cared for me.

Chaca

I guess I spelled his name correctly. He was another great friend, but what made him tick was a mystery to me. One day he refused to play soccer with us. I did not know why, and I pushed him until he agreed. He had a seizure during the game. I was embarrassed afterward for pushing him to play. Rob told me once that Chaca had a seizure while crossing a major road. He was run over and killed. Chaca had a big tender heart. He had a way of blessings those he touched.

School

I struggled in school. Kindergarten was in France. I never had anyone to help me make the transition. Dad was overwhelmed in trying to learn French as well. Mom wanted to help me retain my English when French was what I wanted to get better at. I learned to hear some French in France, but I could not speak it. The next year was in Abidjan, Ivory Coast.

In total, I change languages in school 4 times. We would skip back and forth between English and French all the time. I would slowly catch up in one language. When I would feel caught up to average or slightly better, we would change languages again.

The one area I always thrived in was math. Somehow, math was natural. I could not understand why others struggled. Math was the only subject I felt safe in at school.

Robert Pinkston

Robert was one of my three best friends during my childhood including Kojo and Rob Simrell. Robert and I were very competitive in almost everything. Robert also liked to play keeps with marbles because he was better at marbles. I rarely practiced marbles because I neither wished to lose marbles nor take someone else's. I probably would have become a lot better at the game had we just played for fun.

Robert was an outstanding athlete and a super scholar. He had serious skills. Changing languages in school never bothered Robert. He always found a way to be the top student. When we returned to the United States for sixth grade, my dad asked me if I wanted to skip to seventh. I was a good student in all subjects, but in math I intimidated most of my teachers. I liked the idea a lot. I could move ahead of a grade so that I would not have to compete with Robert. Robert was not the biggest problem. When I would see Robert, I would feel this shame that I brought upon myself because he was better in school than I was except for math class.

God highly esteems Robert and all the Pinkstons. It was Robert who gave me perspective. Robert was the one who said, "There is just something different about the churches that your dad started." For my greatest competitor to admit that my dad was blessed beyond the other missionaries including his dad in church planting was astounding.

I had taken a three-year break between university studies. Nine years after leaving Africa, a student enrolled in the same seminary I attended – it was Robert. Can you imagine randomly walking up to Robert at school? Both of my American friends in Africa ended up enrolling in the same schools that I did. I guess the United States can be a small place after all.

I was married soon after. Robert was not my best friend at my wedding; he was the minister. By then we no longer competed. We were proud of each other's accomplishments.

24 A man of many companions may be ruined, but there is a friend who sticks closer than a brother.

-Proverbs 18:24   (WEB)

Chapter 27

Family Adventures

Bribes

Bribes were one of Africa's biggest problems. It was commonplace for people in power to seek a bribe. Dad had made the choice that he would never pay a bribe. Dad had many clashes with local authorities overpaying bribes.

Tout le Pleut

Tout le Pleut is a small village near the border. The border road was a dirt road. We drove up to the border and dad talked with the officials for about half an hour. Dad came back to the car for a few minutes because he did not want to pay the bribe.

Every few minutes, dad would get out and ask about crossing. The guards continued to refuse to let us pass without paying a bribe. The sun gradually set. We sat in this really hot car just steaming as dad kept negotiating. Finally, the sunset. They did not let us pass and closed the border.

We went back to the village and asked if there was a place to stay. There was not a hotel, but they offered us the nicest house in the village that had space for three people to spend the night. Mom, Patty, and Julia got the chance to spend the night on a bed. Dad and I had to sleep in the car. We cracked the windows to allow the cooler night air to seep in.

We fell asleep. About midnight, I woke up screaming. While I slept, the mosquitoes were biting me. I found out that mosquitoes search for veins just under the skin. When they find a lump, they figure it must be a vein. A mosquito would feel the lump left by the previous mosquito and would bite the bite. I ended up with mosquito bite clumps. I had bite clumps of 7, 8, or more. While sleeping, I had received hundreds of mosquito bites.

Dad asked me to roll up the windows. We went to swatting the mosquitoes in the near dark. We sat there without sleeping the rest of the night in the heat. I couldn't sleep because of the misery. Dad did not dare to go to sleep leaving me as I was.

In the morning, we looked at the ceiling of the car. The paneling was covered with red blood stain splotches. The spots permanently stained the ceiling of the car.

I did not think things were very fair. I wondered if dad had been bit at all. He was not sure if he got bit, but there may have been three or four spots where a mosquito got him.

In the morning, the gals came out from a pleasant stay in the house. We had missed both supper and breakfast because there was nothing to buy. It would take a place like Tout le Pleut (all the rain) to breed that many mosquitoes. In the morning, we returned to the border, and they let us pass without paying the bribe.

Cops

Many of the police just loved to stand by the highway and blow their whistles. They would stop one after another and demand bribes. People had the choice of either paying a bribe or paying a ticket. Oftentimes, the policemen would trump up enough serious charges that paying bribes would seem like an easy escape.

A typical stop would follow a standard plan. As soon as the policeman finished with one motorist, he would blow his whistle at the next car. We would always pass without looking at the policeman pretending that we did not see or hear him. When he would blow his whistle, we would keep driving. Many times, the policeman would let us go seeking easier prey. If the policeman would blow his whistle louder and in many short bursts, dad would stop.

Afraid

When dad would talk to the policeman, he would always tell us to roll the windows up all the way and lock the doors. The heat in a sunbaked car in the tropics was astounding, yet we kept the windows up.

One time, a policeman stopped dad, and he got out of the car. He told us as usual to lock the doors and roll up the windows. After a little bit, the policeman became very agitated. Dad just looked down and said nothing. We were becoming quite concerned. We could not hear the conversation with the windows up. Soon the policeman became hysterical. He was just screaming at dad, and dad was not saying anything. I became afraid that the policeman was going to throw dad in jail or kill him.

After a while, dad just walked away from the upset policeman and returned to the car. We unlocked it so that he could get in. Dad sat down and said, "We can go!" I said, "I thought that he was going to kill you or something!" Dad started to leave like it was no big deal. The whole family chimed in, "He was so mad at you, and you just walked away." Dad responded, "He was not mad at me." The story just made no sense. Dad did not realize that we had not heard the conversation. Dad made it clear, "Once the policeman realized that I was the clergy, he kept screaming, 'Don't send me to hell! Don't send me to hell! Please forgive me! Don't send me to hell!'" God had scared that policeman pretty good that day.

I learned; well, I am not sure what I learned that day. I learned that the appearance of things is not necessarily what is happening. Now as I look back, I see God's hands in ways that I could not see back then. Fear owned me that day because I had not made my choice.

Family Time

On many evenings, we would sit down as a family and play games. We just had the best of times. It did not matter the game. We just loved to spend time together.

Soccer

Still again, we lived in the tropics. The amount of sunlight was very stable. Most days had 12 hours of sun and 12 hours of darkness. It was usually light from 6 am to 6 pm. After school which let out at 4:30 pm, my African friends would come and see me. We would play soccer in the back. Dad would often join us. We would play to exhaustion. When it started getting dark, my friends would leave.

Running

We were a running family; well, some of us were. Mom would run all by herself. I don't blame her. Dad and I ran together nearly every day. Our trek ranged between 1.4 to 4.2 miles depending on where we were living at the time. We took turns winning.

I remember being so tired after having played soccer. Dad was tired too but being tired was never an excuse. After soccer, we would drink some water from the hose. It was not the drinking water that we were too tired to do. Dad was always going to run after playing soccer, so we ran. I could not let him outdo me. How could I claim to be tired when he was a middle-aged man?

Chapter 28
Thinking Styles

Top-Down – Dad

Dad started with one goal. All planning kept in mind the main goal. Dad would not jump to the details on how to solve the main goal. Dad found sub-goals of the main goal. Dad kept building his pyramid of sub-sub goals until he had a complete set of goals.

When Dad went to Ivory Coast now officially La Côte D'Ivoire, he had a plan to reach the whole country to Christ. Ivory Coast had about 12 million inhabitants. Dad's goal was to reach all 12 million. As long as dad was in Ivory Coast he never wavered from that goal.

Bottom-Up – Mom

Mom never really bothered herself with the big picture. Mom loved the tiny details. Mom would write a lot of the correspondence. Many people needed to be contacted. Mom worked hard, doing all the behind-the-scenes things. Mom did so many little things that dad never thought of. If dad thought of them, well it was such agonizing work for him to do it.

Somewhere, dad and mom met in the middle. That somewhere was usually where the details met the goals. Everything was covered because mom and dad were a team.

12 If a man prevails against one who is alone, two shall withstand him; and a threefold cord is not quickly broken.

-Ecclesiastes 4:12 (WEB)

Dad's Hero Missionaries – Top-Down

Dad had to work at getting the answers that he wanted. Dad talked to enough people until he finally came up with a workable solution. Dad would seek out these missionaries with incredible minds that were able to see the big picture. Every chance dad would get, he would talk to them trying to figure out missions. The kind of missions that dad wanted to do was neither taught in seminary nor could be read in books. As far as I could tell, most of the living legend missionaries were top-down thinkers. They just solved bigger problems and reaped bigger rewards. They started with the biggest problem and then worked downward until they found a workable solution.

If you think that I am just rambling theory, then I have successfully begun in the old man's style. These heroes of dad changed the world. They are true heroes of the faith for their ideas reached the ends of the earth. God chose to bless Christianity through these unknown missionaries. God highly esteems them. What these men did will have to wait until later. The setting has begun.

22 Where there is no counsel, plans fail; but in a multitude of counselors, they are established.

-Proverbs 15:22 (WEB)

Dad's Partners in Missions

Several missionaries were eager to let Dad show them how to minister. Dad never wanted to be the boss, he just wanted to give these missionaries a chance to thrive. Whether one starts top-down or bottom-up, both skills need to be learned. Douglass Simrell was one such missionary that learned from dad and thrived. I believe that the Simrells would have been successful anyway because they walked with God and had a humble spirit eager to learn.

Others like Estelle Freeland, Wilma Rodgers, and the Pinkstons just partnered with mom and dad, and also mom and dad with them. They worked side by side with each one using their gifts as able. Each had gifts for what they were called to do, and they worked peacefully together. All these missionaries mentioned here were highly effective and ended up with incredible careers. It was not so much where they started either from the top or bottom, but together there was a whole picture. The whole picture included planning from top to bottom and bottom to top resolving the details.

1 Therefore, let's also, seeing we are surrounded by so great a cloud of witnesses, lay aside every weight and the sin which so easily entangles us, and let's run with perseverance the race that is set before us,

-Hebrews 12:1 (WEB)

Workable Goals – Dad

Dad never saw a way to win Ivory Coast to Christ all at once. I don't think it ever crossed dad's mind that not reaching his goal would have been a failure. Dad would whittle his plan down until he found something workable. Dad would go into villages or communities. Dad had a plan on how to reach the whole village or the whole community for Christ. Dad's plan was trimmed down from trying to reach the whole country to reaching a whole community. The community was not an end, but the community was seen as a new partner in the goal of reaching the country. Dad divided the plan of reaching Ivory Coast for Christ into workable smaller goals.

There is therefore now no condemnation to those who are in Christ Jesus, who don't walk according to the flesh, but according to the Spirit.

-Romans 8:1 (WEB)

Workable Goals – Mom

Mom's workable goal was to make dad a success at whatever God put upon dad's heart. Mom had a belief that God would guide dad in the planning. Mom's goal was to do whatever dad didn't. Dad had a lot of things he could not do well. Mom stayed very busy doing what dad didn't. Mom somehow made all her goals workable. Dad was never caught unprepared for mom made sure of that.

10 Who can find a worthy woman? For her price is far above rubies.11 The heart of her husband trusts in her. He shall have no lack of gain.

-Proverbs 31:10, 11 (WEB)

Workable Goals – Others

My dad's heroes I do not know well. I just know that they had big dreams that they carried out. They will not be discussed much, yet God so greatly used them. I was a kid, and I only saw them every few years. Dad's heroes did not start until they had workable plans. God chose to make those plans echo throughout Christianity.

Many others just looked for a place to start. They wanted to find a place where they could be effective. They so often looked for the smallest effective step and started there. In their faith, they would reach out to the one or ones that God put upon their hearts. These were faithful in doing what God had placed upon their hearts. They started small and worked their way toward bigger things.

17 "He said to him, 'Well done, you good servant! Because you were found faithful with very little, you shall have authority over ten cities.'

-Luke 19:17 (WEB)

Effectiveness of Goals – Bottom up

Targeting a few responsible people can be very effective. Much energy is poured into a few. A lot of discipleship happens when creating the potential for much inner growth. This is the approach that so many missionaries took, and healthy churches were planted.

Sometimes, much time is spent with a few who are responsive socially but never become very spiritually responsive. One of the great difficulties with a few is that it is impossible to foretell who will blossom.

One of my friends had parents who served their whole career as missionaries. It was time to retire. This couple had invested their whole lives in two churches which became rather large. The missionaries had gone into a new area and pioneered work. They did not see their calling as winning the whole country to Christ. They felt called to win two villages to Christ.

At the time that my friend told me about his parents, I was in extreme shock. I thought that every missionary was called to attempt to reach the whole country for Christ. My friend was so proud of his parents' accomplishments as if his parents had fulfilled their calling. Only God sees the end, and He is very pleased with His servants who accomplished their calling.

Ed Pinkston is a very detailed man. He likes to start with the little details. The churches he started blossomed into large thriving churches. God had placed a powerful hand upon the Pinkstons. Ed was not a little man in any sense. He was tall at least compared to dad. His dreams grew big enough to meet the needs of whatever community they reached out to.

Bottom-up is a highly effective way to plant churches. A large portion of missionaries tended to be bottom-up in their way of thinking. Starting bottom-up is a wonderful way to start and can lead to much fruit.

Effectiveness of Goals – Top down

The size of the plan and the likelihood of success do go hand in hand. Those who see little tend to achieve little those who have achieved more have first seen it beforehand. Those who start with a big goal do not necessarily have a big plan. Big goals require comprehensive plans to make them happen. Big goals without an effective strategy lead nowhere. Whether planning starts with little plans or big plans, both are needed. A lot of little plans need to feed big plans, and big plans need to generate smaller plans.

Dad would sometimes drive us crazy. Dad never started a new venture without fully figuring out a plan. Dad never felt obligated to start anything until he was satisfied that he had the best plan possible. Really good options would come and go, and dad stayed neutral. Dad never felt the rush to meet a deadline; yet in all of dad's waiting, we never felt like we missed out on much in the end. I am really comfortable with a very good plan instead of the best plan especially if there does not seem like there is much more to be gained.

One thing that I did learn from watching dad impatiently, life's restrictions would often change with time. With a little waiting, many times whole new possibilities emerge out of the previous impossibilities. When it seems like now or never, the wait was dad's philosophy. If the outcome was not acceptable, dad waited until the outcome became acceptable.

It is not as important whether planning starts top-down or bottom-up as long as the plan is comprehensive. Great planning takes time. Sometimes planning is just waiting for conditions to be ripe.

25 But if we hope for that which we don't see, we wait for it with patience.

-Romans 8:25 (WEB)

Caution of Bottom-Up Planning

Sometimes in bottom-up planning, the planning can be perfect. The pyramid can be complete. All the little plans come together to be one idea. The problem can be that the pyramid is not very big. The thought process is complete for a small task with a small reward. Only a few will benefit when leaving the masses out.

36 But when he saw the multitudes, he was moved with compassion for them because they were harassed and scattered, like sheep without a shepherd.

-Matthew 9:36 (WEB)

Cautions of Top-Down Planning

Top-down planning has its problems as well. Focusing on the masses can make the individual invisible. When the individual realizes that the concern is for numbers, not individuals, then there is no room for individuals in the masses. Planning that reaches the masses still must value the individual.

8 Finally, all of you be like-minded, compassionate, loving as brothers, tenderhearted, courteous,

-1 Peter 3:8 (WEB)

Synergy – Calculating Rewards for Planning

In an effective bottom strategy. Plan A could reach 5 people. Plan B could reach 12 people. Plan C could read 3 people. The logical choice would be to choose plan B while Plan A is perhaps the easiest quick fix. As already mentioned, some choose a quick fix and have reached a plateau way too early. Small thinking chooses one plan in which to invest themselves.

A top-down strategy would have one goal with multiple plans all rolling at the same time. Supervision would be quite different because much of the intimate time would be forfeited to accomplish a bigger plan. With more people to supervise, there are more people to choose from as well. We will get to examples further along in the book.

Running multiple goals all at the same time creates some strange math. Combining the 5, 12, and 3 may make 100. Plans can feed off each other and create more prospects.

Synergy is Not for All

One missionary with whom dad invested a lot of time, became very frustrated with dad. He said, "I see what James does, and it works for him, but it does not work for me. I cannot do what James does." This man was very correct. He spent his whole life longing to accomplish certain kinds of goals. He developed a style suitable to accomplish his goals. He could not develop the skills to have the same kind of success as dad, because his mind was passionate about something else. He never made the transition to dreaming as dad did. Skills have a way of following dreams. This man did not think like dad, nor did he desire to make the transition to thinking like dad.

His goals were different than Dad's. Dad tried to impact as many as he could. This man had much smaller goals that brought him far more satisfaction than bigger goals. The intense focus of passion drives skill acquisition. God has different plans for people. In walking the road called upon, people find the most fabulous treasures.

Loneliness of Comprehensive Planning

If the world is truly dominated by bottom-up thinkers, the tendency is to focus planning on quick easy rewards at the expense of comprehensive planning. Planning often never includes the big picture, just satisfying pictures.

Most people believe that their way of thinking is the most logical of all ways of thinking. The reason people think the way that they think is that it makes the most sense to them. If another way of thinking made more sense, then the new way becomes the only way to think. I imagine paragraphs like this one drives some of you crazy.

Dad found that one of the most common ways of thinking is for a quick immediate result. People gather together to find an immediate answer. They find one and they are very satisfied because the plan is very achievable, and it resolves momentary crises. It takes time to develop the patience to postpone a reward for a bigger reward.

I remember dad several times being overwhelmed by the short-sightedness of others. I remember one time dad returned from a mission meeting; the mission had come up with this wonderful plan that ignored major problems. Dad, in his graciousness, never told us what the problem was. Dad knew that the plan stood no chance of any long-term success.

Dad had a choice; he could have forced his way, which was lonely, or he could have kept relatively silent which was just as lonely.

Comprehensive Planning

Top-down thinking is in no way better than bottom-up or vice versa. The best thinking is not in a style but in comprehensive planning. Without details, a plan is lost is in the sky; with only details, a plan is stuck in a maze. People tend to be consumed in either details or the big picture at the expense of the planning they like least. Being a great thinker is not so much about being smart but working at developing the whole picture.

The Calling

Ultimately, the Designer is the only one who sees the big picture. The Master Architect is weaving together the most amazing plan. People cannot see the whole plan, but the Omnipotent can. The Great Orator speaks a calling that includes only elements of His master plan. The Expert Guide likes to call people down the most terrifying roads. The Great Competitor challenges people to think in styles uncomfortable. The Ultimate Excavator digs through weakness to discover true strength. Blessed is the one who rises to the calling and completes the blueprint engraved by Master Carver.

14 A man shall be satisfied with good by the fruit of his mouth. The work of a man's hands shall be rewarded to him.

-Proverbs 12:14 (WEB)

Chapter 29

Global Missions

Global Missionary Strategy Planning

When we were in Africa, there had not been many or maybe even any global studies on the effectiveness of missionary strategies. Our sending agency decided that they needed to identify the best missionary practices. The goal of one study was to identify the best strategies and then implement them all over Africa.

Visits from Above

One thing that I had a tough time understanding is why we had so many visits from the mission dignitaries. The IMB had 3,000+ missionaries, and mom and dad were only two. I had a hard time understanding why the top brass all seem to know mom and dad. It was only later that I understood.

The idea at the time was somehow to improve the effectiveness of missions. They planned to identify the most effective missionaries and get to know them. Mom and dad were on the top list. The eagerness of the top brass to learn from the best is why they flew halfway around the world to spend time with mom and dad.

> 3 I have filled him with the Spirit of God, in wisdom, and in understanding, and in knowledge, and in all kinds of workmanship,
>
> —Exodus 31:3 (WEB)

The Official Study of Dad's Plan

A committee was appointed to come study dad and other highly effective missionaries found elsewhere in Africa. The first missionaries they studied were mom and dad. Dad laid out his plan before these researchers. Dad's plan was comprehensive. Dad's plan began with the concept of winning all of Ivory Coast to Christ. Dad knew that all was not possible at once, so he devised a strategy that would reach the most people.

It was a really strange day when dad explained the things to the committee. Dad would say something, and a lightbulb came on in their heads. They would be so excited and talk about it for a little while. Once they felt comfortable with the new idea, dad said something else. They were there for about nine hours as dad opened their eyes to his style of missions. It was like they were invited into a whole new world. It was so obvious that after their talk, they had found what they were looking for.

Dad's plan covered these five major categories: ministering to individuals, starting churches, missionary roles, pastoral training, explosion, and transformation.

31 The mouth of the righteous produces wisdom, but the perverse tongue will be cut off.

-Proverbs 10:31 (WEB)

Explosion

Mom and dad knew that they could not start very many churches. Many times, more churches were needed than what they could start. Mom and dad knew exactly what do to; they just did not have the time to start enough churches. For churches to multiply, churches had to start other churches which had to start other churches.

For Ivory Coast to be won for the Lord, time was urgent. Churches could take all the right steps to grow and grow fast. Expanding takes a lot of time and energy. The expansion was happening, and the churches were trying to keep up with the growth. Growing rapidly made one church big, rather than create lots of churches.

Mom and dad's plan was for the churches to be deeply burdened by other communities without Christ. As believers matured, the vision did increase from reaching out to loved ones to reaching out to total strangers. The passion for new churches was central to dad's plan.

Mom and dad started churches, but their plan from the beginning was to develop churches that started other churches. For Ivory Coast to be won for Christ, the locals had to embrace the vision to save the country for Christ. For Ivory Coast to be won for Christ there had to be countless moms and dads starting and growing churches.

32 Everyone therefore who confesses me before men, I will also confess him before my Father who is in heaven.

-Matthew 10:32 (WEB)

Transformation

Dad clearly expressed his ideas to the researchers about his vision for souls. The team was so impressed with dad's captivating spirit. The issue is that they did not know what to do with dad's bigger-than-life personality. The researchers were gifted in taking a plan and writing it into a report.

Their challenge was to place dad into a report. Maybe you can tell by now that my summary's central focus was not on particular events, strategies, success, or troubles, although those things have found their way into the writings. The main focus was trying to capture the things that captured mom and dad. A few people spending a couple of days could not capture the real mom and dad.

I will try to find a few things to sum up mom and dad's passion. Before reading on, can you describe what made mom and dad tick? Humor me, although I will never be able to hear your humor. Humor me anyway; describe why God moved so powerfully within mom and dad. Don't peek at my answer just yet! Describe for yourself their heart and passion.

Here is the 3-for-1 deal. God overflowed mom and dad with love (1). God transformed my parents from being useless to being Godlike (2). God consumed my parents with a vision for others (3). These are the three things that God did for my parents. However, it was a 3-for-1 deal. Mom and dad had to do one thing. They had to let God have his way by surrendering their wills to Him (1). God wanted to do three things for mom and dad, all they had to do was to give God permission.

For each of the three things that God did for mom and dad, here are some of my favorites. Just like mom and dad had been loved by God, they reached out and greeted people warmly (1 – love). Dad spoke to me in the backyard about preaching, "You have to give the Spirit of God room to work" (2 – transformation). Dad's favorite expression was, "God wants to reach through you just as far as God can reach" (3 – calling). Many times, the Africans said that there is no turning back. They tasted the wonderful mysteries of God; they did not want to return. The prize was set before them.

If there is one part of the report that was left out, it was the most important part. God gives his best blessings to the ones who surrender all to Him. God seeks to love, transform, and send the willing soul.

26 For what will it profit a man if he gains the whole world and forfeits his life? Or what will a man give in exchange for his life?

—*Matthew 16:26* *(WEB)*

The Final Report

Mom and dad's plan for the ministry was found to be comprehensive. The final report to the Southern Baptist's Foreign Mission Board about the best way to reach Africa was nearly identical to mom and dad's plan. They chose their report to be like mom and dad's plan.

My memories of the researchers are few. I remember dad explaining everything step by step. These experts were so amazed by mom and dad's plan. You could almost hear their minds open up as they marveled at discoveries.

Many of the most effective missionaries in West Africa collaborated. Mom and dad's report had already included some of the most blessed missionary minds. When the researchers went on and interviewed other missionaries, they were found to be of like mind in so many regards. Mom and dad could not have done what they did without the wisdom of those missionaries. I would talk more about them; except I don't remember anything except the impact they made on mom and dad.

Many of the most effective missionaries practiced the same ideas. The like-minded missionaries collaborated. Little is known about their reports to the researchers, but we do know that the researchers' findings mirrored Mom and dad's report. As mom said to me, "According to the team, Dad had the best and most comprehensive plan. They just rewrote what dad told them."

*19 Wisdom is a strength to the wise man
more than ten rulers who are in a city.*

-Ecclesiastes 7:19 (WEB)

Report Implementation

The Foreign Mission Board received the report with open arms. They realized that they had found some of the most effective strategies in winning Africa for Christ, so they moved on to their next stage. During mission meetings all over Africa, they explained what the most effective missionaries did. Missionaries were taught the best-known strategies on how to win Africa for Christ.

*6 Blessed are those who hunger and thirst for
righteousness, for they shall be filled.*

-Matthew 5:6 (WEB)

Borrowing

During the days of the report, research was in its infancy. Many books had been written, but research reports were rare. Missionaries from other sending groups also desired to know the most effective missionary strategies. There were 2 ways to get such a report. One way was to organize a team of researchers who would travel to the mission field to study missionaries. A second way was to get a copy of a report already in existence. Missionaries all over Africa and the rest of the world quickly latched on to the report. Many missionaries studied what mom and dad did without ever realizing who they were studying.

Since those days, many studies have been made about the most effective mission and church studies. The foundation laid by mom and dad had found its way into many other places seeking the best practices. There is no way of knowing how much of my parent's work eventually found its way into the best church growth strategies, but one thing is certain, mom and dad were central to one of the most influential missionary studies of all time.

42 Give to him who asks you, and don't turn away him who desires to borrow from you.

-Matthew 5:42 (WEB)

The Magnitude of Plan and Action

Dad's plans were comprehensive. Dad's plans were so huge that the researchers, me, and the others, were overwhelmed. Dad stayed with his plans and talked to enough people until he found a way to make them work. Nothing in dad's plan was extremely spectacular except that so many pieces were combined until the plan seemed too grand to work. Once dad started, mom had to race to keep up with her monumental work. Together, they stayed driven to make the plan succeed, and it did.

28 God has set some in the assembly: first apostles, second prophets, third teachers, then miracle workers, then gifts of healings, helps, governments, and various kinds of languages.

-1 Corinthians 12:28 (WEB)

24-hour Challenge

The 24-hour challenge made so much difference. As far as I know, dad was the only missionary to challenge the new believers/seekers to tell their families within 24 hours. Christianity went almost from a spark to a massive fire in a matter of hours. Even the new converts from the new converts took the challenge and ran with it.

16 For God so loved the world, that he gave his one and only Son, that whoever believes in him should not perish, but have eternal life.

- John 3:16 (WEB)

10 to the intent that now through the assembly the manifold wisdom of God might be made known to the principalities and the powers in the heavenly places,

-Ephesians 3:10 (WEB)

The Engines of Planning and the 24-hour Challenge and the Moving of the Spirit

Without massive plans, the explosive growth would have been chaotic. Dad greeted so many people with incredible warmth that the first service usually had 25 or so people. Many more waited to see what would happen after the first meeting. The numbers grew rapidly.

To start 19 churches in three years from nothing in his last tour in Africa, church starts had to be rapid and well organized. Mom had to organize the internal structure of the churches rapidly. The first moment the church could stand on its own, dad appointed a pastor and left.

Dad drew in large numbers and gave them a big vision. The vision became their life's dream instantly to which they dedicated their lives.

The biggest power was the moving of the Spirit. The Africans were not challenged to be great, but to be willing to lose it all so that their families may know the wonders of God. The Spirit of the Living God moves with the greatest power in those who choose to lose most for the Savior. The wonders of God kept the believers racing to be God's hands and feet. God truly did reach through them as far as God could reach. Blessing and honor to God.

1 I command you therefore before God and the Lord Jesus Christ, who will judge the living and the dead at his appearing and his Kingdom: 2 preach the word; be urgent in season and out of season; reprove, rebuke, and exhort with all patience and teaching.

-2 Timothy 4:1, 2 (WEB)

JIM H DARNELL JR

Chapter 30

Miscellaneous Stories

Calling Nigeria

Once in the mid-1970s, dad needed to call Nigeria. Direct dialing was not possible in those days. Dad called the local phone operator in Ivory Coast. The operator could not call Nigeria directly because no phone lines were running from Ivory Coast to Nigeria.

Ivory Coast, which was French-speaking, had its international phone lines connected to France. Nigeria, which was English-speaking, had its phone lines connected to England. Somehow, the operators could not connect directly from France to England at the time, so the connection was routed further – much further. The United States spoke English and Canada spoke French. The connection ended up going from Ivory Coast in West Africa, to France in Europe, to Canada in North America, to the United States still in North America, to England in Europe, and Nigeria in West Africa.

The routing of countries was explained, but the challenges were not explained. The local operator had to call dad back when everything was finalized. A local operator in Nigeria had to call the destination party to inform them of a call that was going to happen at a certain time. Operators in all the various locations had to manually connect the call one by one.

The connection was so long, that it could have just about gone around the world and back if it had been in a straight line. Such a long line made communication difficult. Dad would yell three or four words into the phone then he would stop. The echo came back loud. All was silent for a moment then the echo came back again but not as loud. Two echoes were not enough for the echoes happened about four or five times depending on how loud they yelled. The man on the other end would not yell loud enough for dad to hear very well. After a little bit, dad supposed that the man got the message. All dad needed was a confirmation. Dad heard something so he guessed that was enough.

After each country added its international fees, the call cost $9 a minute. The quality was so poor that less than a third of the time was usable due to the echo. The phone call was a wonder of the day.

This belongs here just because there is not a better place. The call was difficult, expensive, and very low quality. The call is just like life. Life seems so haphazard and poor quality at times. From heaven's perspective, nothing could be clearer or more rewarding.

10 All the paths of Yahweh are loving kindness and truth to such as keep his covenant and his testimonies.

-Psalms 25:10 (WEB)

The Crippled Man

Dad went to start a church in a village. There was an old man in the village who had been crippled from polio for 12 years. This man wanted to walk again. When the old man moved around, he had to drag his useless legs. He wanted to be healed by God.

Needing a miracle was not the only issue. Neither dad nor the man spoke the same language. To communicate, they needed not 1, but 2 translators. The old man asked dad for a miracle. Translator 1 expressed the need for a miracle in another language. An interpreter who understood that language knew

French and spoke to dad. For an answer dad spoke French to translator 2 who spoke to translator 1 in a strange language. Translator 1 spoke back to the old man in a second strange language.

Dad did not know what was getting through and what was being warped. Both men wanted something. The old man wanted to be healed and dad wanted the man to be saved. Dad had never seen anyone healed from an incurable disease. Dad only knew one thing to do; he just trusted the Healer.

Dad led the man through prayer by the interpreters. As soon as the man finished the prayer, he chose to put his newfound faith to the test. The old man hopped up and began to walk around.

I was away at boarding school during the miracle. When I went to the church, the members were still overawed by the miracle. They were bragging about what God had done for the man.

It was not long before this man was summoned. His former village wanted him to be their chief. Once this man could function again, his services were needed.

From a heavenly perspective, this man was sent ahead to bring salvation to his home village. We never heard the outcome, but we do not need to. God had a plan, and he put his servant just where He wanted him. We do not know the rest of the plan, but God was excited about it.

> 6 But Peter said, "I have no silver or gold, but what I have, that I give you. In the name of Jesus Christ of Nazareth, get up and walk!"
>
> -Acts 3:6 (WEB)

5 Beloved, you do a faithful work in whatever you accomplish for those who are brothers and strangers.

-3 John 5:5 (WEB)

Ed and the Sea

I used to like to push the boundaries; I thought real men had to logically figure out how close they could get to death without dying. In some of those episodes, it dawned upon me, "I could have died!"

In Ivory Coast, we had huge waves. The waves were 20 feet high on a calm day. These waves would crash near the shore or even on the shore in places. The waves threw huge tongues of water upon the shore. The water would ferociously rush back to sea clinging to anything in its way.

We only went to the beach once with the Pinkstons. Once was enough for the incredible loyalty shown to me by Ed Pinkston. Ed (Robert's dad) only wanted to watch us swim and me, in particular.

The sea posed two great dangers. Sometimes, the receding water would flush the sand out from underneath the feet. Without footing, the rushing waters would haul even the strongest person out to sea. This danger is not the one that humbled me for I stayed close enough to the shore where the undertow was not quite as strong.

A second danger came from the crashing waves. The crashing of the waves was so intense that it would dig a trench. At one moment, the water would be belly high with an invisible trench directly in front. Once, the receding waters stole my footing forcing me to take a step forward. I stepped into the trench and disappeared. It was strange to take a small step into nothing. Suddenly, I was spinning in the waters, and I could not figure out which way was up. I realized then that I was out of my league. While I was spinning in the water, I quickly concluded that I did not have the skills nor the strength to save my life. An arm wrapped around me; it was Ed's. Once things calmed down. We were way out at sea. I realized that we were too far out to swim back. Even far from shore, the sea was vicious for huge swells would rush toward us. On the smaller waves, Ed just tucked us under the wall of water to let it pass us by. Once a really big wave came, Ed tried to climb up the wave. The wave caught Ed and me, and we raced back toward the land. The water was carrying us so fast that when we made contact with land, we were moving so fast, we ran to keep our balance.

I was still confident that I could be careful enough if I just played it right. I went right out and played some more. Before long, I lost my footing and stepped into the well dug by a wave. Once again, I could not figure out which way was up. The same arm wrapped around me again. By the time the waters calmed back down, we were way out at sea. I anticipated Ed catching a huge wave, and we once more were driven back to shore.

I decided on a few things that day. I decided that the ocean was not as fun as it used to be. I realized that I was powerless to save myself. Sometimes, being brave was nothing more than stupid. I could not see beyond my embarrassment to see Ed. I wished that I had thanked him – twice.

The Pine Tree

There was a pine tree in our front yard. The tree had grown extremely tall. One day, I decided to see how high I could climb. There were a lot of branches, and the climbing was easy. As I went up, the branches became smaller. These thinner branches bent easily. I made sure that I stepped up against the trunk and held on as close to the central column as possible. I was proud of myself for being smart for I was figuring out the boundaries of safety.

After a while, I pulled on a branch, and it broke. It was okay because I still had two feet and one hand firmly attached. I found out that this tree was taller than the others. I had a clear view of the huge lagoon that surrounded our neighborhood. I saw the far shore and much more. I wondered if I saw nearly the whole city of a million to a million and a half.

I kept climbing. After a while, I realize that I was safe as long as I had one hand and one foot firmly always set. The branches grew weaker, and I still went higher. After a while, the branches were so flimsy that I could only remove one foot or one hand at a time without snapping a branch. I kept going further. I thought I could go a little bit more and still be safe.

Now and then, I would look straight down. I did not like the view. The house looked little and falling would have surely meant death. I could not think about that, I had to think about how I could live.

At some point, I woke up. I knew I could climb higher, but what was there to gain? I proved I could be brave and think logically in the face of death. I knew I could not go much higher. Even the trunk was swaying with each movement. I decided that logically cheating death for no gain was illogical. I came down. I don't think that I ever told that story until now. Now is not even the writing of the book. Now is the editing. Enough of this story, but there was one more that bothered me more.

The Waterfall

Mom and dad were always taking us to zoos and other places of great interest around the world. Once in Africa, we were taken to a waterfall.

They kept telling us the falls were 10 stories high. Mom and dad said that Patty and I could go around to the top of the falls. Patty and I ran around and arrived at the very flat top. The falls were not a whole lot more than a trickle. The water was not deep at all. The water was deep enough to make a splash, but that was about it. Not only was the water extremely shallow, but the flow was a baby's crawl. The water leading to the falls was unimpressive.

From the top, we could look down. The water below was a crystal-clear pond about a yard deep with huge boulders poking out. Being the expert that I thought I was on how safe things were, I decided quickly that going over the edge was not safe. Landing on a boulder was certain death, and landing in shallow water was not any better.

I decided that I had to cross the stream. All I had to do was take one step in the water. The water was so shallow that it was not even going to rise beyond the sole of my tennis shoe. The only thing that concerned me was the moss I knew that grew on rocks that were underwater. I figured that if I balanced myself perfectly, I would not slip.

Something else caught my eye. The rock was so barren that nothing seemed to grow except one little wimpy-looking plant. It had one thin stalk rising that seemed like a joke.

I told Patty, "Come on, let's go." Patty stayed still. I moved forwards without her. I stepped perfectly. There was not anyway I could tumble because the pressure from my step was completely downwards. I was so correct; I did not tumble. I also found out that the slipperiness of the invisible moss was a real concern. Immediately, the slow current started pushing my foot toward the cliff. I was perfectly balanced on one foot proving that I was so right and deathly wrong at the same time. I immediately dropped to all fours hoping to stop myself, but I could not. I was flowing along at a baby's craw, and there was nothing to stop me.

I cried out to Patty, "Stop me!" She did not move. There was nothing she could do. I was pushed over that insignificant plant that I looked down upon, but that little plant had my full attention. I grabbed the puny stalk, and I was surprised. It held. I reached a foot over to dry land, and then I easily found safety.

I was humbled that day. I did everything right, and all so wrong. I was pushed over by a trickle, and the frailest of plants saved my life. That was the last time, I risked my life for fun. I have spent the rest of my life still developing the bravery to stand for what I believe in which has helped me in sports. I have spent my life studying the things that I believe God wants me to stand for. As you can tell, this book is about a full commitment to what counts. Since then, I have worked on bravery and how to avoid stupidity.

Chapter 31

First Churches

The Wonders of Firsts

First churches always had a special place in mom and dad's hearts. First churches are by definition the first churches that my parents started on different continents. Mom and dad never talked favorites; there are just some practical reasons why first churches kept the limelight.

Columbus

Mom and dad started a church in Columbus, Ohio. They spent 4 years establishing that church. Those four years are by far the second-longest period as a minister at any church during my lifetime. For ministers with long ministries, 4 years is so short. They could have sunk their whole lives in Ohio and had been perfectly content. I just remember mom and dad being happy. It was also the first church that they ever started. They were full of wonder and excitement. Mom had friends she kept in contact with all her life from that church. The first church start was a shining beacon reminder of all the good that can come out of starting churches.

There were a few other pastors who prayed together for their ministries. The Associational Director gave dad a foundation in church planting work. This foundation was exactly what Dad needed skill-wise to launch churches. The men and dad had such an incredible experience. They would lay on the ground before the Lord seeking His face. The revival atmosphere had come!

Koumassi

Koumassi is a neighborhood of Abidjan. As a family, we spent more time in Koumassi than in any other church in Africa. The only place mom and dad could find to meet was a bar. Bars were unused on Sunday mornings making them available.

The church was full of strong leadership. Koumassi was the place appointed by God to touch Ivory Coast. Dad drew so many men out of the church to be pastors. This church became the founding rock on which much of its ministry was built.

I almost did not include the first church section, because I do not want churches to feel ranked. Some churches might read this and feel less. To you, the other churches, there is not this sense of unfinished business. Mom and dad were able to pass along the best of what God had given them. Mom and dad were fully satisfied with the blessing passed on. Because you seemed to fully receive the blessing, mom and dad were incredibly pleased.

Cayenne

The first church that was planted in South America was in Cayenne French Guiana. Outside of Cayenne was a small town/suburb called Montjoli (pretty mount). That church will be the focus of the next section of the book.

Chapter 32
Final African Stories

Church Count in Africa

Dad never said the important things twice to me. I wonder how much I have forgotten that would have made my life much richer. When we left Africa for the last time, dad said that they had started 25 Yoruba churches and 20 French-speaking churches. When dad said it, he made sure it was just the two of us for dad hated talking about his accomplishments.

3 But when you do merciful deeds, don't let your left hand know what your right hand does,

-Matthew 6:3 (WEB)

Updated Church Count

Technically, mom and dad only started 45 churches in Africa, but that is not true. One of mom and dad's missionary bosses said, "I can find 50 men to keep a church going, but where can I find one to start churches?" Even among missionaries who were trained to start churches, those who could were rare.

Dad would scatter his church starts all over Ivory Coast. In the middle of almost nowhere, they would start a church. Shortly after the church started, another one would start beside it. One denomination faced the incredible difficulty of starting a church from nothing. Their solution was to go wherever mom and dad started a church and draw members away. Often mom and dad's one church became two. Our parent's church starts were so blessed that there were enough believers for two churches to thrive. I don't know how many of these double churches starts, but dad named at least five while we were still living in Africa.

Do these offshoots count? Without mom and dad starting new work, there were not have been either of the two. Mom and dad do not need credit for what they did. However, writing this book has given me some new perspectives.

14 "Glory to God in the highest, on earth peace, good will toward men."

-Luke 2:14 (WEB)

Hello!

African love and kindness never ceased to amaze me. During our stay in Africa, twice we returned from a one-year stay in the United States. Upon our return to the mission field, we would get flooded with African visitors. It was not uncommon for someone to travel all day or for several days.

Many walked in the heat all day because they did not have the money for a bus fare. They would ring the doorbell. We would answer. They just wanted to greet us all with a strong handshake and kind words. They never stayed for more than a minute or two. They would tell us, "Bye" and shake our hands once more. They made their intentions very clear. They did not want to take up our time; they just wanted to greet us.

Many would tell us about the sacrifice made in seeing us. Here are some of the most typical stories. "I walked all day because I did not have money for a bus fare." "I traveled two days to see you." "I did not eat for three days so I would have enough money to come see you." Many did not count the cost to come see us. They loved us greatly.

Dad knew that many also came with a request. The request was always the same, "I do not have enough money to get home. Can you give me enough to take the bus home?" They never asked for food or drink. It was embarrassing enough for them to ask, but they had to come.

What makes these stories so beautiful is who came. Most of these people who came, we did not know who they were. I always ask dad, "Who was it?" "I don't know" came the most common reply. They knew who they were was not important. They had come to salvation through mom and dad's ministry. What they knew is a family had left the comforts of America to tell them about Jesus. They had chosen to be as much like mom and dad as possible. Mom and dad were the best example of what God looked like to them.

The Africans usually came one at a time. Their count increased into the hundreds just because they had to tell us how much they loved us. God's love had grown strong in them.

20 All the brothers greet you. Greet one another with a holy kiss.

-1 Corinthians 16:20 (WEB)

Farewell

I love this part. The French language has changed some over time. There is a term for goodbye in French called, "Adieu." The letter "a" means in this case, "till." "Dieu" is the French word for "God." "Adieu" meant "till God." The expression means "Bye forever till we see each other again in heaven!" Now the term is used more freely than "Bye."

When we left Ivory Coast, we went to their airport. This was before airport regulations became strict. About 300 Africans walked us out to the plane. They each took their turn to tell us,

"Adieu." They only had to say it five times. We had to say it 300 hundred times.

They just started singing choruses. They did not just pick any chorus. We all knew the same songs, but they made sure they picked songs that were in English also. They wanted to honor God and honor us who left our homeland to tell them about Jesus. We always sang in French at churches even when we knew the same song in English. On that particular day, there were five of us who sang in English.

When it was time for the plane to leave, we boarded. We saw them no more – not a single soul either present or not. Many wonders of our lives had come to an end.

22 The Lord Jesus Christ be with your spirit. Grace be with you. Amen.

-2 Timothy 4:22 (WEB)

Bibles

I know the story of Africa is done, but the goodbye is not where I want to end it. I would rather think about Bibles. Bibles can have a different meaning to the poor and uneducated in lands of persecution.

Many Africans would carry their Bibles wherever they went. If they walked three miles or seven, they carried their Bibles. Their Bibles meant so much to them that many would not leave home without the Scriptures.

Most in mom and dad's ministry, had made their choice. They chose to love God with all their hearts. Carrying a Bible was their declaration of love and loyalty to God. They had risked losing their families, homes, jobs, spouses to be, and anything they called a future. They had chosen to risk it all for they had found God's great love. Carrying the Bible was telling the world of God's love for them.

Bibles were also a witnessing tool. People would notice the Bible, and it would give a natural opportunity for them to talk about the wonders of God. The Bibles would give opportunities to talk to strangers about God.

Mom and dad sold Bibles for about 50 cents or $1. The Bibles did not cost the Africans much, but to many the price was extreme. Some only had one or two sets of clothing. Both shirts had holes in them. The pants had ripped knees and other tears. They could have saved for new clothes which were quite needed. Some went hungry until they could afford a Bible. Having a Bible was not an option, but a necessity of faith.

We talked about being poor in the lands of persecution, but we missed a subject. We have not talked about uneducated. Why does someone who cannot read at all need a Bible? Why miss eating for 2 or 3 days in the baking sun to purchase an unreadable book. To them, the answer is obvious. Sometimes, they could find someone who would read to them even if they were strangers. Mostly, they had to carry the Bibles because they loved God.

16 Every Scripture is God-breathed and profitable for teaching, for reproof, for correction, and for instruction in righteousness, 17 that each person who belongs to God may be complete, thoroughly equipped for every good work.

-2 Timothy 3:16, 17 (WEB)

John Mills

John Mills was mom and dad's boss in Africa. He had traveled all over West Africa supervising hundreds of missionaries at a time. John Mill's job was to guide missionaries to become all they could. "The Darnells are the best missionaries I have ever seen!" John also said, "I have supervised 1,000 missionaries, and the Darnells are the best." John realized that God had blessed mom and dad's work uniquely.

24 A man of many companions may be ruined, but there is a friend who sticks closer than a brother.

-Proverbs 18:24   (WEB)

Chapter 33

French Guiana

Furlough

Before we left Africa, mom and dad started getting requests to preach in the United States. Mom and dad would visit churches regardless of size. By the time they had been in the United States for three days, they were already booked up for the year. Mom and dad had speaking engagements in churches up to five a week. When the Sundays ran out, they added any day of the week. Mom and dad liked going around preaching.

Many people responded. Decisions were made in abundance. Some received salvation. Many rededicated their lives. A lot caught a fire that God could use them, too. Far more people than dad could count dedicated themselves to be foreign missionaries. Dad spent a lot of time with pastors wrestling with becoming missionaries. (In the years to come, the Foreign Mission Board kept informing my parents when another new missionary appointee declared mom and dad as the main reason for becoming missionaries.) God moved people.

2 that their hearts may be comforted, they being knit together in love, and gaining all riches of the full assurance of understanding, that they may know the mystery of God, both of the Father and of Christ, 3 in whom all the treasures of wisdom and knowledge are hidden.

-Colossians 2:2, 3 (WEB)

1½ Years after Africa

Dad called me on the phone, "The Foreign Mission Board had just called. They found a place for us to serve in South America. There is a French-speaking country called French Guiana. There are not any America Missionaries there. In helping us understand what do to, we would like to know your opinion."

3 praying together for us also, that God may open to us a door for the word, to speak the mystery of Christ, for which I am also in bonds,

-Colossians 4:3 (WEB)

French Guiana

Two years after leaving Africa for the last time, mom and dad were on a new journey to French Guiana. Mom and dad were refreshed. They were ready to start again.

My memories of French Guiana are few, mainly because I did not make many. Some kids run away from home. For Patty and me, it was the other way around. Mom and Dad ran away from home. Patty and I continued to stay in the United States while mom, dad, and Julia went off to South America. I made two journeys to French Guiana: one during my college days and then about two years after graduating.

New Strategy

You would think that after perfecting a strategy that worked great in Africa, they would try the same strategy. Mom and dad had not had a church home since they started the church in Ohio. They were ready for a different approach. They decided that they needed the main church which would branch out and start other churches. God does not have just one best way.

They Started a Church in their Home

Mom and dad started a church service in their home. At first, they started meeting in their living room. Soon their dining room was needed as well. Before we rush along with the story anymore, let's describe the setting.

The Climate

Cayenne, French Guiana, just like Abidjan Ivory Coast was 6 degrees off the equator. That is about as close to the equator as possible without actually being there. There were some weather differences as well between the two continents.

In Africa, they had rainy and dry seasons. Africa had huge thunderstorms. A tree in our yard and our hot water tank had both been hit at some point by lightning. The lightning in Africa when it struck close was the loudest thing I have ever heard. In French Guiana, they had the rainy season and the season when it rained. All that rain was without lightning except the rarest flicker.

Both cities: Abidjan and Cayenne were coastal cities carved out of the jungle. The humidity from the ocean was always extreme. The humidity was so thick that evaporation did not work very well. Since sweat did not dry very well, the body's natural cooling off mechanism was not too effective.

The House

The house was perfectly built for the tropics. The living room and dining room were the middle third of the house. The middle of the house did not have outside walls and windows.

Rather the wall was replaced by two layers. The first layer was a set of removable sliding glass doors. The second layer of protection was an ornate metal grill.

Mom and dad permanently took off the glass doors. To keep people out, the bars could be locked. The living room and dining area was like a huge, covered porch, surrounded by a house on two of the four sides. Since the house did not have air-conditioning, the designed cooling system was a breeze and fans.

Beyond the dining was a cement patio. In the front, there was a carport which was a sheltered cement slab, but an open area on two sides. The yard was uncommonly big in all directions. The yard in front of the house was open like a small field. The sides had seven mango trees scattered around. All these details were designed by the Master Planner for times to come. The house could not have been designed any better for its appointed destiny.

Poverty

Poverty can be destructive to societies. The extremely poor lack so much to escape poverty. The poor often lack education, social skills, and motivation to make life different. There is a toughness that comes with being poor. The poor too often find themselves at the bottom of the social scale. The most helpless are often bullied. The issue is not about being charming or having social etiquette; the most helpless can easily find themselves harassed at every turn.

4 Rescue the weak and needy. Deliver them out of the hand of the wicked."

-Psalms 82:4 (WEB)

Haitians

Haitians came from a country called Haiti which is part of an island in the Caribbean. The land mass was too small to support all the inhabitants. Those who stayed in Haiti often did not have much access to education and career opportunities. Haitians have spread all over the Americas to find new opportunities. About as many Haitians lived in French Guiana as natives.

Haitians Responded to Church

While there were many different ethnic groups in French Guiana, it was a segment of the Haitians that were willing to come to church. Some of the Haitians were culturally trained that it was okay to go to church. When Haitians first started going to church, it was very often not for religious reasons. Since the stigma for going to church was not negative for some, curiosity, and other factors we will discuss drew them in.

Mom and dad could have tried to reach out to other people groups and had success. However, there was more work that could be done among the Haitians than mom and dad could do in a lifetime. Mom and dad just worked where there was a harvest.

16 You didn't choose me, but I chose you and appointed you, that you should go and bear fruit, and that your fruit should remain; that whatever you will ask of the Father in my name, he may give it to you.

- John 15:16 (WEB)

JIM H DARNELL JR

Chapter 34

Marriage & Divorce

Separated Families

Many times, the men would leave their families behind in Haiti. The common plan was for the man of the house to discover a better life, then the families would join him. The problem with the plan is that every place had plenty of poor. While many of these men took the adventure, a large portion could not find opportunities for a better life. They neither had the means to support their families nor the money to pay their way to come.

The more unfortunate found themselves stuck apart. The man did not want to go back as a failure and the families could not come to him. The families found themselves in separate worlds.

> *51 Do you think that I have come to give peace in the earth? I tell you, no, but rather division.*
>
> *-Luke 12:51 (WEB)*

Unfaithfulness

After a while, unfaithfulness often struck. It may have been the man, woman, or both who were unfaithful at first. Lack of faithfulness was a huge problem for these families where the man was gone for extended amounts of time.

When I traveled to French Guiana, I heard similar stories like this one too often. A broken-hearted man would say that he had found a job. He kept as little as possible and was sending the rest back to Haiti. He was so proud that he could finally support his wife and kids. He could not send much because he did not make much, but it was enough for his kids not to go hungry. He was paying an ultimate sacrifice for them. He was all alone but loyal. After a while, he started hearing rumors from his friends that his wife was not being faithful. His money was not only supporting his wife and kids, but also another man. Men often found that they were supporting unfaithful wives.

Likewise, a man would depart on a new adventure to a strange land. The new adventure was another woman. The argument often goes like this, "What my wife does not know won't hurt her." In his mind, he can still take care of his wife and kids while pretending all is under control until he hears these words, "I am pregnant!" The man then had a new woman with a baby coming on the way. He realized that his wife would hear the news. The man was struck with a terrible choice. How could he be faithful to the woman and kids left behind when they know he is a sorry excuse for a man? How could he leave his new woman and their baby?

Others tell stories like this one. I left, and we both had affairs. The marriage was over because neither one committed to sticking it out during the exceedingly difficult financial times.

Families that are driven apart by extreme hardships often find themselves with new crises. They were far apart with little hope at the time of being reunited. New romantic flings were the easy way out for the moment rupturing their vows one to another.

52 For from now on, there will be five in one house divided, three against two, and two against three. 53 They will be divided, father against son, and son against father; mother against daughter, and daughter against her mother; mother-in-law against her daughter-in-law, and daughter-in-law against her mother-in-law."

-Luke 12:52, 53 (WEB)

The Divorce Curse

I have a tough time understanding the divorce curse. I can explain the logic, but the curse itself makes no sense. Divorce has a unique curse that nothing else in society bears. This curse is especially strong among Haitians.

The curse is that anyone who divorces is a moral and spiritual failure. No shame is bigger in society than being divorced. Sleeping together outside of marriage does not carry the same shame that divorce does. The curse stems from the Bible, but it is not Biblical for it is misinterpreted. Scripture expresses that church leaders should be the husband of one wife. When divorce happens, neither one is seen as acceptable to church leadership and the divorcee is seen as second-class members. Divorce under the curse nearly ends ministry within the church.

The curse is hard to understand, but it often only applies to divorce and not faithlessness. For those who "shop around" before marriage and have many sexual partners, the curse does not apply. Once the right one is found, marriage can eventually happen without any penalty. Unless the couple divorces, the curse does not apply. They can be considered highly esteemed leaders of the church for being married only once.

I don't follow the logic of the curse. Those who intentionally chose to do things the wrong way cannot be penalized under the curse, while those who intentionally chose to be responsible can only fail. The divorced are often rejected by the church and it teaches others that immorality is good.

5 I left you in Crete for this reason, that you would set in order the things that were lacking and appoint elders in every city, as I directed you, 6 if anyone is blameless, the husband of one wife, having children who believe, who are not accused of loose or unruly behavior.

-Titus 1:5, 6 (WEB)

Dad Understood the Divorce Curse

Dad never expressed an opinion to me about his mother's divorce. Dad did not like talking about the pains of his childhood. His dad had found romance with other women. Eventually, dad's parents divorced. However, dad felt the sting of the divorce curse and dad was just a kid. The United States in those days often strongly believed in the divorce curse. What dad and his mom did know best are the sorrows of living under the divorce curse when his mom had been faithful.

8 He said to them, "Moses, because of the hardness of your hearts, allowed you to divorce your wives, but from the beginning it has not been so.

-Matthew 19:8 (WEB)

The Faithful

Many Haitians were faithful to their spouses. They lived their faith and were very loyal one to another. These are incredible people. Against all odds, they stayed together. In the face of enormous trials, they found a way to make it past the misery and thrive. They showed that success was possible, and they lit the way for others to follow. These are people to be highly esteemed.

7 A righteous man walks in integrity. Blessed are his children after him.

-Proverbs 20:7 (WEB)

JIM H DARNELL JR

Chapter 35

Broken No More

Prospects for New Churches

Those available to be the founding blocks for new churches were the broken and the needy whom most were not walking with God. They have tasted the sting of failure. The churches drew from the unchurched pool of broken, needy, and angry.

> *10 For the Son of Man came to seek and to save that which was lost."*
>
> *-Luke 19:10 (WEB)*

The Need to Hit Bottom

As long as people feel like they can write their own rules for life, they do not need God. When people realize that they lack the strength to change then they need God. Those, who realize that they cannot find their way without God, know they must look for the Savior.

> *3 For we were also once foolish, disobedient, deceived, serving various lusts and pleasures, living in malice and envy, hateful, and hating one another.*
>
> *-Titus 3:3 (WEB)*

Amazed by the Newcomers

I was amazed by some of the attitudes of those who first came to church without yet having a personal faith in God. Some came to church strutting like a rooster. Others came angry. Many came with a wall protecting their wounds; they tried to look cool and hide their sorrows. Others just came broken without much hope of a future.

22 That which makes a man to be desired is his kindness. A poor man is better than a liar. 23 The fear of Yahweh leads to life, then contentment; he rests and will not be touched by trouble.

-Proverbs 19:22, 23 (WEB)

Dad's Message to those Who Felt so Unlovable

Dad preached certain themes over and over again. Dad was wanting to give the Spirit room to work by presenting God's view. For things to change, it had to start with what God wanted to do.

Dad preached, "God loves you just the way you are!" These people who felt so unlovable realized that God loves them in their present mess. The Spirit of God said, "I love you just like you are!" The loveless realized that God's love was waiting to flood their starved souls. God's love drew them to Salvation. In their moment of crisis, they found their answer in the love of the Savior.

8 But God commends his own love toward us, in that while we were yet sinners, Christ died for us.

-Romans 5:8 (WEB)

The Message to those who Felt Hopeless

The hopeless had two great sources from mom and dad's journeys. Those seeking a better life yet could not find it. The others were those who messed up life so terribly that there was no hope of fixing it. The message was simple, "God has a plan for your life." To those who could not find their way, the burden to find the way was God's. He said, "Let me show you the way; I have wonderful plans for your life." For those who crashed their lives through failures, God said, "Let me fix your life; I have wonderful plans for your future." They could feel the fix upon their hearts just waiting to hop in.

> 10 For if while we were enemies, we were reconciled to God through the death of his Son, much more, being reconciled, we will be saved by his life. 11 Not only so, but we also rejoice in God through our Lord Jesus Christ, through whom we have now received the reconciliation.
>
> -Romans 5:10, 11 (WEB)

The Message to those under the Divorce Curse

Dad's message to them was the same with a little extra, "God loves you just the way you are. God has a plan for your life that He is very excited about. He wants you to be His hands and feet. Get married!" For the first time, many of the divorced felt like they had a place in church. God spoke, "I love you just the way you are. You feel like your life is ruined through a divorce, but I have a wonderful plan! Let me honor your present relationship!"

> 20 The law came in that the trespass might abound; but where sin abounded, grace abounded more exceedingly; 21 that as sin reigned in death, even so grace might reign through righteousness to

eternal life through Jesus Christ our Lord.

-Romans 5:20, 21 (WEB)

Amazed by God's Touch

Like night turns into day so quickly, some changed from hitting bottom to reaching heaven. The mean agitated spirits became gentle. God took the bitterness and anxiety far away; instead, they were sloshing in God's love. God so changed them, they no longer needed to be mean. The broken became so full that they were glowing with love.

14 who gave himself for us, that he might redeem us from all iniquity, and purify for himself a people for his own possession, zealous for good works.

-Titus 2:14 (WEB)

The Marrying Preacher

Dad developed a new nickname as the "marrying preacher"! People wanted to get married. Some had lived together so long that their kids were adults. The church eagerly celebrated all marriages. The past failures were not the important issues, all that mattered was the present blessing of God upon their lives.

There was a great need to have God's blessing upon their relationships. People were coming all the time for God's incredible love and honor. Marriages just happened and happened a lot.

2 But, because of sexual immoralities, let each man have his own wife, and let each woman have her own husband.

-1 Corinthians 7:2 (WEB)

Chapter 36

Music

The Band

Dad had prayed that God would send Him a way to attract more people. One Sunday morning, a couple of cars showed up at mom and dad's house. They were a band needing a sponsor. Dad did not know anything about them, but he let them stay and play that Sunday. How could he send them away before seeing if they were the answer to prayer?

They had one Sunday before they blew an amplifier. They needed a lot of equipment that they could not afford. Dad was surprised at how much money was needed to be spent on this band. They were always needing something else.

3 Praise him with the sounding of the trumpet! Praise him with harp and lyre

-Psalms 150:3 (WEB)

Music

To avoid copyright issues, the band wrote their own music. Writing their music is an overstatement. Most could not read music but play by ear. They could hear a new piece, and within moments join right in.

Over the years, they had produced several CDs. To be the lead singer of the song, each had to write his own piece. They took their music seriously and practiced a lot.

21 You therefore who teach another, don't you teach yourself? You who preach that a man shouldn't steal, do you steal?

-Romans 2:21 (WEB)

Spirituality of the Band

The band had organized because they loved music and longed to be musicians. The band had base, melody, and accompaniment guitars. They had drums and a keyboard. In addition, there were several singers.

Just because they were musicians, it did not mean that they were spiritually mature. Some did not have much of a spiritual foundation. The music was drawing them to Christ.

Doisy really stood out. He was not in the business of making himself great. He played because he loved God, and this gentleman was just captured in the moment as he played to God.

6 Let everything that has breath praise Yah! Praise Yah!

-Psalms 150:6 (WEB)

The Band became Highly Loved by the Church

As time passed, the band became part of the backbone of the church. Whatever the church did, they were eager to sing along with the band for God's glory. The band had to be there.

4 Make a joyful noise to Yahweh, all the earth!
Burst out and sing for joy, yes, sing praises!

-Psalms 98:4 (WEB)

Julia's Picture

It had been a couple of years since I had seen Julia. One day, I received a picture in the mail. My parents assumed that I would know that it was Julia. She no longer looked like a kid; she looked like a young adult. The picture that I was staring at was a young pretty girl. I knew who it had to be, but I am not sure I would have recognized her on the street. I was filled with two feelings. I was saddened by missing Julia's growing up years. I was also very happy that she had turned into such a pretty young woman.

Julia and the Clarinet

When Julia was in the sixth grade, she learned to play the clarinet. In Africa, she wanted to play it for the band. The band played by ear and Julia could only read music. Most of the musicians could not help her. There was one member who on occasion would write music for Julia to play.

The weather was harsh on the corks underneath the finger valves. No one in French Guiana worked on clarinets. Julia did not have anyone to service her instrument. The valves started leaking air as the cork rotted away.

The vendors in French Guiana sold many postcards. One picture was printed on a cork sheet. Julia tried to fix her clarinet herself. She bought a cork postcard and cut it into the exact circle size. Julia learned how to fix her clarinet. I wonder how long it took her to get it right the first time. Did Julia fear that she would never get to play her instrument again?

Julia in the Band

Julia stood out as a band member. She was a cute petite white girl. She drew the attraction of lots of guys. After she played, Julia was hit upon by many guys. It was funny watching guys trying to outdo each other to impress Julia. Julia knew why she played. She played to bring people to Christ.

Chapter 37

Legalism

Legalism was a part of the Christian culture. So many lives were being destroyed by alcohol and sex outside of marriage, and legalism was a religious response. Legalism is a focus on right and wrong. The only way to stop wrong from a legalistic viewpoint is to find and punish wrong. Legalists are very zealous to judge the wrongs.

9 For the zeal of your house consumes me. The reproaches of those who reproach you have fallen on me.

-Psalms 69:9 (WEB)

The Bar Patrol

One member of the church came up with the idea to stop drinking and immorality. The plan was to stop the drinking and lewd behavior that went on in bars by some of the church members. His solution was to have a "bar patrol" with badges. They were to be given the authority by the church to investigate bars. For the patrollers, the badges would honor the Christians' entrance into the bar.

This church member had worked many into a tizzy. When Sunday came along, they were ready to go to war against the Christians who went to bars. At first, dad stood alone in the fight against the bar patrol for the people did not understand the value of grace.

2 For I testify about them that they have a zeal for God, but not according to knowledge.

-Romans 10:2   (WEB)

Pierre

Pierre was a member of the band. He was the leader. Pierre had a great voice and a charismatic personality. Pierre was also stubborn. He was so stubborn in fact, he just would not back down.

Pierre also had another habit. He listened to dad and tried to understand these strange concepts coming out of dad's mouth. For being such a stubborn man, it seemed odd that he was so willing to listen and understand views that were different than his own.

1 My son, if you will receive my words, and store up my commandments within you,2 so as to turn your ear to wisdom, and apply your heart to understanding;

-Proverbs 2:1, 2   (WEB)

Pierre's Response to the Bar Patrol

Pierre let dad fight an unhappy crowd for a while. A large majority of the church members were tired of the few who disgraced the rest. There were several who were quite upset that dad refused to make the members accountable for their behavior. Pierre listened to dad's arguments until he understood them. Pierre continued to stay silent until he figured out how to express the arguments himself.

Once Pierre was ready, it was his turn. Dad just needed to step aside for then it was Pierre's fight. Pierre argued with them until they gave up. They knew that Pierre would never back down except to dad, and they were on the same side. Many were not happy, but they accepted defeat in the proposal to have a bar patrol.

9 Don't grumble, brothers, against one another, so that you won't be judged. Behold, the judge stands at the door.

- James 5:9 (WEB)

Dad's and Pierre's Message

Wrongs need to be addressed but fighting evil should not be the focus of the church. A patrol may keep more members in line, but it does not change the hearts. True change comes from God's grace. Change does not need to happen because of fear of getting caught; change happens when God touches lives.

37 Don't judge, and you won't be judged. Don't condemn, and you won't be condemned. Set free, and you will be set free.

-Luke 6:37 (WEB)

Legalism Zeal

Legalism often generates extreme fervor while accomplishing little good and lots of harm. Legalism is very strong at pointing fingers in two directions. Legalism always points fingers at those who falter, "Look how bad they are!" Legalism points fingers at self, "Look how great I am for not being like them."

It would only seem natural that legalistic people would lead very moral lives, and they often do for a while. The problem is that passion wilts legalistic zeal. The voice within cries, "I want it! I want it! I want it!" Cravings drown out legalistic passion. When legalistic strength finally wears out, moral failure happens unless God intervenes.

Legalism is a weaker passion than lust. The fingers that pointed out the moral failures of others and personal greatness eventually point to other reasons. When legalism is overcome with passion, the fingers point to self as being the failure.

Therefore, you are without excuse, O man, whoever you are who judge. For in that which you judge another, you condemn yourself. For you who judge practice the same things.

-Romans 2:1 (WEB)

The Power of Grace

Grace is God's gift to the guilty. To the weak, grace is strength. To the shamed, grace overlooks failure. Grace is God's answer to all problems. To the hopeless, grace is a fountain of hope. To the bitter, grace is joy. To the angry, grace is peace. To the jealous, grace is owning the wealth of heaven. To the judgmental, grace longs for others to be transformed. Grace is ever different yet always the same. Grace replaces the worst part of man with the best God can give.

9 who saved us and called us with a holy calling, not according to our works, but according to his own purpose and grace, which was given to us in Christ Jesus before times eternal,

-2 Timothy 1:9 (WEB)

God's Response to the Judgmental Attitude

Most strongly opposed dad that day. It was a lonely day for dad and Pierre, but God used their voices. After the arguing was over God spoke clearly. To some, He spoke sooner than others, but God eventually made His plan clear.

God spoke love. "You remember when you were so ashamed of yourself, you couldn't stand life? Do you remember what I did? I was not pressing my finger on your guilt to destroy you, but to redeem you. I invited you to come to my love and forgiveness. I pointed out your faults and invited you to come to my grace." God reminded them of his love, but that was not all.

God also spoke of fear. "Do you think you can enjoy my love and grace, and be judgmental to the rest that I love? Do you think that I will let my love flow all over you and through you while you refuse to love others? You must choose now if you want to keep the warmth of my love within you for, I am about to yank it from you."

God spoke moral tests. "Will you take your eyes off these things you judge and place them on me?" For those who did not listen, God increased the intensity of the test. In the face of passion, their strength weakened. "Will you take your eyes off these things and look at me?" As the test increased so did the blame, excuses, and frustrations. For some who kept their eyes off God and upon the sins of others, many experienced moral failures. The very thing that they judged so strongly they committed. For others, God aborted the test right before the moral catastrophe. They were left realizing they would have failed except for the grace of God. God spoke tests to demonstrate that judgment only leads to moral failure; they need their eyes upon God.

The very one who stirred up the whole church to have a bar patrol continued to stir the church up to be legalists. This young man fell in love one day. If anybody knew better than to live in immorality with a woman, it would be he. He staked his whole reputation on being pure and fighting drinking and premarital sex. The church noticed that his girl-friend's belly was growing at a very predictable rate. You could say that she was just getting fatter, but only one part of her body was growing. Her arms and legs stayed the same size, but her middle was swelling big-time. One Sunday, lots of church members accused him of a double standard – premarital sex is not okay for others, but it was for him. That day brought him great shame. His main problem was neither the premarital sex nor the pregnancy. His problem was that he chose to take the place of God and judge everyone. He discovered that he was not strong enough to be God, and the shame of the pregnancy was the evidence. This young man approached dad and said, "I am so sorry for trying to start the bar patrol and judging everyone else." He found that being judgmental was a terrific way to crash his own life.

God spoke love, fear, and tests. God reminded people of the love they received when they deserved judgment. God spoke fear to those who judged rather than passing on the same love. God tested them displaying how truly weak judgmental people really are and how far they have wandered from grace. God speaks to judgmental people calling them back to grace.

7 that being justified by his grace, we might be made heirs according to the hope of eternal life.

-Titus 3:7 (WEB)

Negativity

Growth comes through tests. Many of those who became Christians had spent their lives being negative. Had the church been stacked with mature Christians, the negativity would not have been such an issue. The church was full of new Christians who had yet to face tests about negativity.

10 For, "He who would love life and see good days, let him keep his tongue from evil and his lips from speaking deceit.

-1 Peter 3:10 (WEB)

Dad's Response

Many in the church were continuing to stir up the negative, judgmental attitude. All these bitter people wanted to do was fight. Dad finally had enough of the negativity. Dad made a shocking announcement, "I resign as pastor of this church. You can have this church and I will start a new one." The church members, responded as one, "Oh pastor that is not what we want!" They realized that they were causing dad so much pain.

1 Better is a dry morsel with quietness, than a house full of feasting with strife.

-Proverbs 17:1 (WEB)

Transformation

From that moment on, a new spirit fell over the church. The test opened their eyes to the destructiveness of negativity. The church took its eyes off problems and focused its gaze on God. A sweet spirit fell over the church for a new spirit had come.

15 All the days of the afflicted are wretched, but one who has a cheerful heart enjoys a continual feast.

-Proverbs 15:15 (WEB)

JIM H DARNELL JR

Chapter 38

Church Growth

T he church kept growing and growing. People from all over were desiring to come. All the time new people came. As the church grew accommodations were necessary.

Living Room, Dining Room, and Back Porch

After a while, the church increased in number to the point where the living room and dining room were not sufficient to hold all who desired to attend. The cement floor continued out of the house to create a porch. Without the glass doors, and the bars being opened, the whole area was usable. However, the sun blazed upon the back porch. The solution was simple; they covered the back porch. The living room, dining room, and back porch combined to make the auditorium for worship.

Long before my first visit, this area became too small. My first visit was 2½ years after their arrival. The church continued to grow, and new accommodations were needed.

Dining Room, Back Porch, and Back Yard

To enlarge the worship area, the plan changed. The back porch and dining room became the stage area. The roof over the back porch was extended over the yard. Before long, they had the same problem again; the worship area was too small. As dad once expressed, "We just kept extending and extending the roof." During my two visits, the roof that extended out past the porch would cover over 300 people. The house was perfect to accommodate large growth in worship.

47 praising God and having favor with all the people. The Lord added to the assembly day by day those who were being saved.

—Acts 2:47 (WEB)

Sunday School

Sunday School set-up was very differently because there were not any classrooms. The carport was ideal for a children's class. Since the base was cement enabling folding tables to sit firmly. Mom always found many activities for the children to do that required sitting at tables. Sunday School classes were spread all over. One class met in the living room. Other classes met under the seven mango trees, and the covered back porch. Every shaded area on the property ended up being a refuge for Sunday School classes.

The Choir

Many wanted a choir. Finding people who could sing was not the problem. Fining those who could read music was a problem. No one was found who could read music and sing. Mom ended up being choir director. Mom was willing to do whatever, or not do whatever. Mom could read music and be able to guide in singing harmony.

Sometimes, the choir would sing songs that the musicians did not know. The choir and band did not practice together. On Sunday, the band members listened until they identified their parts. Within ten seconds most were playing. Now and then, when the band members were not sure what was coming next, they would stop and listen.

Discipleship Training

The church members desired to grow closer to God. They were wanting training in specialized subjects. The only one they knew that could organize it was mom. No one else had enough depth to lead them into a deeper spiritual journey so they asked her to lead. I remember the people being ecstatic with the new spiritual discoveries. Every Sunday night, when the study time ended, the people kept talking about wonderful discoveries of the Christian faith. Over and over, they encountered the Living One.

Drama Director

Mom organized a church play. You would think that a church play was not that big of a deal. Many had such marginal roles in society that having a part in a play was a very big deal. The reason so many had no experience and so few skills is that the church drew in those who society often cast out.

Woman's Role in the Church

After a while, many of the men came to a crisis of belief. They had always believed that men should be in charge of the church. Mom has/had served as choir director, Sunday School director, drama director, and now discipleship training director. Mom did not serve because she was power-hungry. Mom served because she was the only one who had the necessary skills. The men in the church wanted her to serve but couldn't because they did not have the skills. It was funny listening to the men. At least the men were honest and open about their crisis of belief. One day the main men in the church had to openly discuss the pressing issue on their minds. The men unanimously said, "I have always believed that women should never be leaders of the church, but Mrs. Darnell is better than all of us." They had that conversation within her hearing range which was their way of affirming mom. She got a lot of giggles out of that.

Friendly Competition

Dad once sponsored a competition while I was there on who could bring the most visitors to the church. Dad was so excited about this competition. You could tell that dad was eager for more to know Christ. Dad's focus was on seeing people make peace, and for the church to develop a vision for the lost.

On the day of the competition, many people brought visitors. The church was about twice the normal size. One man took the competition very seriously and brought 40 people. A woman brought 8. Many brought others.

5 But you be sober in all things, suffer hardship, do the work of an evangelist, and fulfill your ministry.

-2 Timothy 4:5 (WEB)

No Youth

One of the problems that plagued the church was a lack of youth. There were lots of adults and lots of kids, but no youth. Eventually, the children would have grown into youth, but dad had a passion for the youth of the country.

The Year of the Youth

French Guiana was about to officially celebrate the year of the youth. The government was going to honor the youth in a big way. Dad decided that the year of the youth was the best time to launch a youth program. Dad wanted to reach the youth in the biggest way possible.

My First Visit

Dad and I have vastly different skillsets especially when I was younger. I wanted so much to develop the kind of skills that dad had. Anytime, dad would open his mouth about church work, I would pay great attention. When dad opened his mouth about the year of his youth, I was flabbergasted. His plans were so huge.

One day, when dad and I were alone, dad brought up the subject of starting a youth program. Dad said, "I have been thinking about how to start a youth program for two years, but I don't quite have it figured out yet." I knew that dad had pondered on this subject for countless hours.

Dad expressed all the things that dad wanted to do, but he did not quite know how he wanted to do them. As I listened, the how was obvious, but I did not say anything at first. After dad had finished revealing his dreams about starting a youth program, I knew all the answers that dad could not find.

I never did tell my dad what he should do. It was his plan and his dream. I just asked dad a question to which he had an obvious answer. Dad expressed his answer. When dad was satisfied, I asked a follow-up question. Dad answered this question easily. I kept asking dad questions that he answered one after another. When the fuzziness was gone, I had run out of questions. Dad was satisfied that the plan was complete. Dad was ready to start preparing for the year of the youth.

I have thought about this episode many times. A half-hour of my time had made a significant impact on the ministry. For the first time, dad and I were on equal terms. God had done a marvelous thing. God had given dad a part of the answer and the other part to me. God had reserved a portion of the ministry just for me.

Flyers

Dad had printed far more flyers than he needed. Dad had rented a pilot to fly over the area and drop "flyers" if you get the pun. After the flyers were dropped, the whole town knew that we were starting a youth group.

Radio Advertisement

Dad placed radio ads on the station that were popular with the youth. Dad wanted the youth to know that the beginning of the youth program was a big deal. Through the flyers and the advertisements, they were given a location and a time.

A Float

Dad entered a float into the Year of the Youth parade. Dad rented a flatbed truck for the event. The band and their equipment played on the flatbed. Many church members walked along beside the truck handing out flyers.

The Bandstand

A bandstand was built in the front yard. Normally, the front yard was reserved for parking. For this event, the bandstand was decorated in a very appealing way. Everything was set for the big event.

Day One

The time finally came for the first youth meeting. On that day, 125 youth showed up. Julia and the band played on the bandstand. The rest of the front yard was set aside for games. They had volleyball nets and lots of other activities. Snacks were served. Dad gave a devotional, and many responded.

Remember also your Creator in the days of your youth, before the evil days came, and the years draw near, when you will say, " I have no pleasure in them;"

-Ecclesiastes 12:1 (WEB)

Youth Leaders

Recruiting leadership for the youth was part of the challenge. Recruiting people to start the youth program was a new venture for the leadership. Most had not been in church very long. They were trying to get Christianity figured out for themselves, now they were asked to lead. Most of them had never had much leadership experience. Even mom and dad did not know how many were going to show up. It was an exciting adventure for the youth and the leadership as well.

After the first meeting, dad turned over the youth program into mom's hands. It was mom's responsibility to organize and develop the youth program. The new teachers needed mom's constant attention at first until they were confident on how to lead the youth.

14 But you remain in the things which you have learned and have been assured of, knowing from whom you have learned them.

-2 Timothy 3:14 (WEB)

The Youth Group

When I arrived for my second visit not too long after the Year of the Youth, the youth group was active and strong. The regular attendance was 40+ with more than that which came regularly. What my parents did not want, was to draw the youth from other churches. Sure enough, the youth that came knew little to nothing about God, salvation, and the Bible.

4 I have no greater joy than this: to hear about my children walking in truth.

-3 John 4 (WEB)

JIM H DARNELL JR

Chapter 39

A Few More Stories

Tennis

Dad played a lot of tennis. As long as his eyesight permitted, Dad continued to play tennis regularly. In French Guiana, the tournaments were on the weekends, so dad did not participate in the tournaments where they earned their rankings. Tennis was a huge deal, and there were so many talented players. These players were nationally ranked in the French system. They would get so mad at times losing to an unranked player. One day, dad got a ranking in the mail. The locals solved their problems by ranking dad somehow.

Hitchhikers

One of my first memories of church in the United States was not inspiring, yet, I guess, somehow it is. I don't know how to define them. I was in the second grade while we were on furlough. It was in the United States. We had just come out of a church. An older man was standing in the church parking lot. He wore an old tie and a worn-out suit jacket. He asked us for a ride home from church. Dad said, "No!" I was really upset at dad for stranding the man. It was not just me. The whole family was upset at dad. The next morning, mom and dad looked at the newspaper. On the front page was a picture of the same man in the same clothes. The article said that someone had given the man a ride from church, and he had killed his hosts. I realized then that God had protected us. On that day,

God had protected us from a most vicious yet unintimidating-looking stranger. The event was my first introduction to the idea that God's paths do not always look right.

When I went to French Guiana, the story of the killing hitchhiker still burned in my mind. Dad and I were in the car together one day in French Guiana while we were returning to the house. A man was walking by the side of the road. Dad wanted me to stop and pick the man up. I asked, "Aren't you afraid of strangers?" Dad's answer was simple, "No, I pick them up every chance I get!" From then on, I picked up all the strangers that I could who were walking beside the road in French Guiana.

12 For now we see in a mirror, dimly, but then face to face. Now I know in part, but then I will know fully, even as I was also fully known. 13 But now faith, hope, and love remain—these three. The greatest of these is love.

-1 Corinthians 13:12, 13 (WEB)

Julia's Schooling

Julia had a price to pay if mom and dad were to stay missionaries. Julia needed to study English to prepare her for university studies in the United States. There were no English-speaking schools in French Guiana. Julia was left with four choices. If she wanted, mom and dad would have returned to the United States so that she could continue her studies in English. Another option was to live with family in the United States. She could also go to a boarding school in a different country maybe not too far away. She could be homeschooled which was also an option. Julia had several choices to continue her studies in English.

Darlene Flaming

Darlene Flaming was a missionary journeywoman. She came with the express purpose of being Julia's tutor and working with the English church. Darlene had been an outstanding student, and she passed on the learning skills to Julia. For two years, Julia studied with Darlene.

One of the two churches that worked closely with Mom and dad which they did not start was the English-speaking church. Darlene was faithful to the congregation. She was always taking them out to visit prospects. The church was so pleased with the leadership and gentle spirit.

15 Give diligence to present yourself approved by God, a workman who doesn't need to be ashamed, properly handling the Word of Truth.

-2 Timothy 2:15 (WEB)

The English Church

On my second trip to French Guiana, I worked with the English church. Some of the members of the church whined. Their negative spirits drove me crazy. I asked God to show me how to deal with it. Whenever someone complained to me, I said, "I heard that before." About a month later, they knew exactly what I was going to say when they whined. After they complained to each other, they started to mouth my words with me. After that, anytime someone complained, the joke was, "I heard that before." The joke ended the whining spirit in the church.

The church had dropped down to about 12 in attendance. When I left six months later, the church was running 65. Before long, mom told me that the English church had over 300 in attendance.

23 But refuse foolish and ignorant questionings, knowing that they generate strife. 24 The Lord's servant must not quarrel, but be gentle toward all, able to teach, patient, 25 in gentleness correcting those who oppose him: perhaps God may give them repentance leading to a full knowledge of the truth,

-2 Timothy 2:23-25 (WEB)

Pastoral Training

As in Africa, dad offered pastoral training classes. Whoever wished to attend could. Up to this point, only one church had been started by mom and dad. They had brought the English church into their home as their own. Pastoral training was effective in preparing leaders for the ministry.

20 Now in a large house there are not only vessels of gold and of silver, but also of wood and of clay. Some are for honor, and some for dishonor. 21 If anyone therefore purges himself from these, he will be a vessel for honor, sanctified, and suitable for the master's use, prepared for every good work.

-2 Timothy 2:20, 21 (WEB)

Chapter 40

The Spreading of the Gospel

More Churches

After a while, God burdened the hearts of several men to start churches. Starting a church is a matter of perspective. When a leader developed a following within the church at our home, he often sought to start a new church with his faithful. Dad never had a leader approach him and ask if he could take a portion of the church with him. This man would promote his new church heavily within the congregation that mom and dad had started, and then he tried to draw away as many members as possible.

Dad would always talk to these men. Dad had nurtured many of them along from infancy in Christ to be ready to lead a church. Dad always sent them out with his blessing. The departure was not treated as a split but as a departure. On the last Sunday before their departure, the new church was sent off.

These departures significantly shrunk the attendance at the home church. One group left with 80 people. Before long, more had come and taken their place.

I do not know for sure how many new churches had been started through these departures. At one point, I was told of five. This count of five was in their first 6 of 20 years in French Guiana. All sorts of other denominations sprung up.

Some of the men had roots in other denominations. As a result, the new churches often identified themselves with other denominations.

God planned to start churches all over. While mom and dad never envisioned themselves starting churches for many different denominations, God had a plan. God gave mom and dad many bonus churches.

40 For whoever is not against us is on our side.

-Mark 9:40 (WEB)

Turnover

Many of the new members of the church did not stay long. They had come to French Guiana seeking their fortune. They found themselves in church; and before long, they made peace with God. God took the broken and turned them into spiritual giants. They left with a vision of God changing their world for Christ. These rough people had turned gentle. Immaturity was replaced with compassionate leadership skills.

When they left, they left on a mission. God had a plan with exciting new ministries. The useless came, left, and were used mightily by God elsewhere.

Some men left because they were estranged from their wives. The men were compelled to return home. God was in the business of restoring many marriages. Many had been transformed from rebellious to gentle leaders through God's touch. God opened doors in other countries for so many of them. They often left for family or business reasons, but God was on missions sending out spiritual giants.

9 Let love be without hypocrisy. Abhor that which is evil. Cling to that which is good. 10 In

*love of the brothers be tenderly affectionate to one
another; in honor preferring one another;*

-Romans 12:9, 10 (WEB)

Jobs in Other Towns in French Guiana

Now and then, a family would move away to another town in French Guiana. A town that was void of any protestant churches. Dad and the family saw the move as a wonderful opportunity to start another church. Once again, dad would be in the church starting a business.

6 having gifts differing according to the grace that was given to us: if prophecy, let's prophesy according to the proportion of our faith; 7 or service, let's give ourselves to service; or he who teaches, to his teaching; 8 or he who exhorts, to his exhorting; he who gives, let him do it with generosity; he who rules, with diligence; he who shows mercy, with cheerfulness.

-Romans 12:6-8 (WEB)

The Story Goes On

I would like to tell you more, but I can't. I never returned to French Guiana after my second trip. A few more useful tidbits can be shared. I could tell you more, I suppose under different conditions. If God had wished for me to tell more of the story, I would have had another opportunity to visit French Guiana. I also know that I have forgotten far more than I remember. Still, I must be content. I told the most important part of my memories. I have passed along the treasures given to me, and I have done my part. I was only there for the beginning of the ministry in French Guiana. Far more wonderful things happened that I cannot begin to express. Mom and dad found the ministry in French Guiana to be extremely satisfying. French Guiana was a place of healing and incredible joy.

Chapter 41

Ministry Summary

Number of Churches in All

I do not know how to count the churches that were started directly or indirectly by my parents. I will try. Dad said that they started 62 churches in all. 2 more (the English church +1) worked so closely with dad that claimed each other. At least five more churches were sent out of the main church in French Guiana. Other denominational leaders in Africa went around placing churches next to the one's mom and dad started. They intended to start and grow churches in large part of the blessing God gave to mom and dad. There are also countless hundreds of grandchildren churches that have mushroomed all over Ivory Coast, French Guiana, and many other countries around the world. The estimated church count by continent that mom and dad physically started themselves were: North America: 1, Africa: 45, South America: 16. There were at least 10 offshoots and two more where dad was loved like the founder plus all the grandchildren churches.

It is difficult writing this section because dad never liked talking numbers. Dad did not like it when pastors assembled and tried to outdo one another by bragging about numbers. Still, how can a book be written on our parent's ministry without a church count included?

My mind wanders back to the churches that moved out of the original congregation in French Guiana. Mom and dad just passed along God's blessing. The church count is important for a reason, but not too important. The church count is a physical description of the incredible blessings God gave mom and dad, but the blessings are not to be held on to. As mom and dad sent the splits away with a blessing, so are the church counts sent away into God's hands. Mom and Dad were God's hands and feet, but the churches were not theirs. The churches were all God's. To Him be the Glory forever.

12 But we beg you, brothers, to know those who labor among you, and are over you in the Lord, and admonish you, 13 and to respect and honor them in love for their work's sake. Be at peace among yourselves.

-1 Thessalonians 5:12, 13 (WEB)

Other Accomplishments

There were other things God did through mom and dad other than starting churches. Dad was the general contractor in building three houses for missionaries, and the contractor in building churches. Dad taught pastoral training classes preparing men who otherwise had no opportunity to prepare for the ministry. Mom and dad trained other missionaries on how to be effective. They spoke in countless churches about missions in the United States fanning the flame in which multitudes of decisions were made. Dad mentored many who sought to be missionaries. They organized ministry to international students connecting them with Christian families. Dad utilized a correspondence course that trained many in the fundamentals of the faith and brought many to Salvation. Dad helped organize a movie that radically changed ministry all over Africa. Dad also did a lot of the basic chores and errands needed to accomplish so much.

Mom

Mom chose to do whatever dad didn't. She did much of the

countless correspondence that needed to be done which was so much harder before the days of electronics. Mom did all the little things that churches have multitudes of committees to do. Mom did whatever dad did not do. Dad would often start projects and then let mom grow them like Sunday Schools and a youth program. Mom recruited teachers for Sunday School and trained them. Often, she had to assist many teachers at once, jumping between them and keeping classes going. Mom initiated her choir, church plays, and discipleship training programs. Even the men leaders of the church came to acknowledge that she was more talented than them all. Mom's greatest accomplishment was trusting God to lead her husband through. She did not try to control him, she just believed in the One who gave her James.

Missions 101

In the early days of mom and dad's ministry, studies of best practices on the mission field were scarce to non-existent. A team was appointed to study the most effective missionaries in Africa. The team was amazed by dad's ministry vision. Dad's plan was found comprehensive and consistent with the kind of things done by the other most effective missionaries. The team's final report mirrored dad's. The report was taught to the rest of the SBC (IMB) missionaries in Africa. Many ministers all around the world studied and incorporated the ideas into their ministries especially since such studies were scarce. As dad greatly borrowed, so did others, and new reports came in their season. While the full impact of dad's plan is not known, it is believed that dad significantly contributed to the ideas that shape the ministerial and mission best practices taught in seminaries to this day.

The only thing that the report was not able to capture was dad's love and personal vision. God has a plan to reach through people as far as God can reach. God calls people to risk it all.

God so loves those who give Him their all that there is no turning back.

Chapter 42

A King's Reward

Dad's Dream

Two years before dad died, the family went to see mom and dad. Our son, Jonathan was a competitive swimmer and needed some pool time during vacation. James (our oldest son), Jonathan, dad, and I went to the pool. We tried to give dad some good memories during those agonizing days. We took dad into the back door right by the pool and sat him down. Before long dad fell asleep and slept through practice. We woke dad up when it was time to leave. Dad could not focus on where to put his feet, so James and I walked on either side supporting him. At one point while walking, dad fell asleep. James yelled, "Grandpa is asleep!" We woke him back up, and from that point on, James and I carried dad out the rest of the way. All dad had to do was move his feet. Dad moved his feet less and less until he was again asleep. We had to wake dad again to put him in the car.

I did not know what to do. Dad's body was in a crisis for something bad was happening. We had taken dad to the pool knowing any little thing could kill him. Still, we had chosen to give dad one more memory to make his last days more bearable.

Once dad was in the car, I had to drive somewhere. The logical place was the hospital. Dad also said, "If anything ever happens to me, don't take me to the hospital." Dad was very willing to live, yet so ready to die. Both the house and the hospital were in the same direction. I drove while we talked about it. Finally, I had to make a choice. If I took dad to the hospital, he would be upset and would have felt like I betrayed him. I could not take him to the hospital, so I drove home.

James and I drug dad into the bedroom and laid him diagonally across the bed. Mom did not say much for she had made her choice to honor dad's request. Mom said little and was always very kind, "You did the right thing. It is what dad wanted. Right now, he isn't living." Dad laid there all day asleep. Dad kept twitching and jerking for something violent was happening to him as we had never seen before.

I can't imagine the burden Jonathan would have borne if dad died because he chose to watch Jonathan swim. Jonathan was 11 at the time. Jonathan had made his choice, too. Jonathan had agreed to help dad live even if it accelerated dad's death. I made it clear to Jonathan that even as kids we must make hard choices. I was proud of him for being willing at great risk to make dad feel like a member of the family.

About eight hours later at sundown, dad woke up. He seemed back to normal. Dad had a story to tell. Dad said, "I fought a great battle!" Dad continued, "I earned a king's reward!" Dad thought for a moment and continued, "I know I did."

That is all dad said. Those were some of the very last sentences I heard dad speak. Dad spent the last two years of his life confined in a body with about the only thing that worked well was pain. Mixed in with all the troubles were the memories of that great battle.

While dad may have been dreaming, the battle was very real spiritually. God so kindly spoke to dad, "You have fought a great battle and have won a king's reward!" God did more than speak. God has put this man in a lifelike dream. This broken man was alive once more fighting on God's behalf with great power. Dad had fought against the mighty powers of darkness victoriously. Before dad awoke, the battle had ended, and dad was triumphant. God honored dad with a great reward for winning the battle. God gave dad a chance to reenact his life that was fully pleasing. God had made His opinion clear on whom He loves and esteems.

While this may seem trivial, it is not to me. If I had taken dad to the hospital, they would have likely given him medicines that would have interrupted dad's dream quest. Everything within me screamed to take dad to the hospital. The choice to go home was to honor dad and throw dad's life into God's hands. By choosing to take dad home, it preserved God's blessing upon dad. God honored the choice and spared us many sorrows. Neither James nor Jonathan objected to taking dad home.

Sometimes I wonder if the dream was more for dad or us. God made sure that I was there to record the feat. We saw dad's ravaged body and only saw a shadow of times past; God saw a conquering king. I know for me, I needed to hear that making the most of every moment was worth living.

6 For I am already being offered, and the time of my departure has come. 7 I have fought the good fight. I have finished the course. I have kept the faith. 8 From now on, the crown of righteousness is stored up for me, which the Lord, the righteous judge, will give to me on that day; and not to me only, but also to all those who have loved his appearing.

-2 Timothy 4:6-8 (WEB)

A Useless Bullfrog Story

Even in the most modern tropical cities, some animals could still be found. One such animal is a bullfrog. If there is any stagnant water, frogs will lay countless strings of eggs. The strings are a perfect set of evenly spaced pearls. Given enough time, the pearls would hatch and tadpoles would come out. After a while, the tadpoles would start to look like tiny frogs. The cute little creatures would turn into huge ugly bullfrogs.

Dad did not always take the advice given to him. In French Guiana, dad was told, "Do not stay in the bathtub when you pull the plug!" Dad continued to take baths and he ignored the advice. One day, dad reached over to pull the plug. Dad had one leg on each side of the drain hole as he reached down. Dad pulled the plug. As soon as the plug was pulled, out jumped a huge bullfrog. The bullfrog did not just land anywhere. As already said, Dad had one leg on each side of the bathtub. With one great surprising leap, the frog came out of nowhere to cover dad's privates. Dad never would tell me how he jumped. He did say this, "I never again stayed in the bathtub when I pulled the plug."

Yopougongare

On November 9, 1975, under a tree in the neighborhood called Yopougongare, Abidjan, Ivory Coast, a church was born. I asked Dad, "Do you think anyone will come?" Dad answered, "Yes, they will come." Come, they did. On that first Sunday, there were about 40 people. The church kept growing. Six years later, they numbered about 600 who had risked their lives for Christ. They were challenged not to live as half Christians by only being willing to lose it all for Christ. The challenge was to be full Christians by seeking to redeem the moments and let God reach through them as far as God could reach. The church went forward and took the challenge. God has been reaching through them even to the ends of the earth. They have sent out hundreds of missionaries all over the world and have grown to be one of the largest protestant churches in all the world if not the largest.

Farewell

French Christian have one more saying, "Que la paix de Dieu soit avec vous d'aujourd'hui à jamais!" "May the peace of God be with you from today till never!"

16 Now may the Lord of peace himself give you peace at all times in all ways. The Lord be with you all.

-2 Thessalonians 3:16   (WEB)

SCROLL OF HONOR

James H. Darnell

FOR SERVICE RENDERED
in the cause of Christ and Southern Baptists in world missions

IVORY COAST 1969-1982

FRENCH GUIANA 1982-1998

ON BEHALF OF THE INTERNATIONAL MISSION BOARD OF THE SOUTHERN BAPTIST CONVENTION

CHAIRMAN OF THE BOARD   December 31, 1998   PRESIDENT

Made in the USA
Coppell, TX
03 January 2023

10318699R00184